Books are to be returned on or before
the last date below.

**3 – DAY
LOAN**

LIBREX-

# Natural Heritage

It has become more and more accepted that nature conservation is not possible without taking into account human activities. Thus an integrated approach to both the natural and cultural heritage is being encouraged and developed. Gathering a number of distinguished authors with diverse backgrounds (from a religious leader to academics to conservation scientists), the book investigates the relationship between human beings and nature; between nature and culture.

Looking at nature as 'heritage' of the human race is a recognition both of the tremendous impacts (both positive and negative) that human activities have had on the natural environment, as well as the acceptance of human responsibility for managing our planet in a sustainable and sensitive manner.

The texts included examine this interface between human beings and nature in specific places (from the Everglades in Florida and Mont St Michel in Atlantic France, to the UK, Europe and the Mediterranean), as well as on a theoretical basis, and in the context of the international biodiversity conventions.

This book was previously published as a special issue of *International Journal of Heritage Studies*.

**Peter Howard** founded the *International Journal of Heritage Studies* and edited it for 12 volumes. He taught landscape and heritage issues at Plymouth University, and is now Visiting Professor of Cultural Landscape, Bournemouth University, and International Officer, Landscape Research Group. He is author of *Heritage: Management, Interpretation, Identity*.

**Thymio Papayannis**, an architect and planner, contributed to the establishment of WWF Greece, the Society for the Protection of Prespa and the MedWet Initiative. He recently founded Med-INA (Mediterranean Institute for Nature and Anthropos) and is its director. He has written numerous articles and five books on environmental and planning issues.

# Natural Heritage

At the Interface of Nature and Culture

Edited by Peter Howard and Thymio Papayannis

Routledge
Taylor & Francis Group

LONDON AND NEW YORK

First published 2007 by Routledge
2 Park Square, Milton Park, Abingdon, Oxon, OX14 4RN

Simultaneously published in the USA and Canada
by Routledge
270 Madison Avenue, New York, NY 10016

*Routledge is an imprint of the Taylor & Francis Group, an informa business*

© 2007 Taylor & Francis

Typeset in  Minion by Genesis Typesetting Ltd, Rochester, Kent

*British Library Cataloguing in Publication Data*
A catalogue record for this book is available from the British Library

ISBN 10: 0-415-44142-0 (hbk)
ISBN 13: 978-0-415-44142-1 (hbk)

# Contents

# Contributors

Bartholomew, Archbishop of Constantinople and Ecumenical Patriarch.

Salvatore Arico is Programme Specialist in Biodiversity at the Division of Ecological and Earth Sciences, UNESCO, having also worked in the secretariat of the Convention on Biological Diversity. He has interests in reconciling cultural and natural heritage issues, especially in a marine context.

Peter Bridgewater has been Secretary General of the Ramsar Convention on Wetlands, and has had a long involvement in natural and cultural heritage issues. Published works include (with Salvatore Arico) 'Conserving and Managing Biodiversity Sustainably: The Roles of Science and Society', *Natural Resources Forum* 26 (2002): 245–48, and (with Celia Bridgewater) 'Is there a Future for Cultural Landscapes?', in *The New Dimensions of the European Landscape*, edited by R. H. G. Jongman (Springer, 2004).

Giorgos Catsadorakis studied biology at the University of Athens and his PhD thesis (1990) was on bird ecology and conservation. He has authored seven books on nature conservation and many scientific papers and articles. He is a freelance consultant on issues of ornithology, habitat management and conservation and environmental monitoring and interpretation. For the past 10 years he has been working as a Scientific Advisor to WWF Greece. In 2001 he received the Goldman Environmental Prize for Europe, for his contribution to the conservation of Prespa and the designation of the Prespa Park, the first trans-boundary protected area in the Balkans.

Alan Dyer is co-director of the Centre for Sustainable Futures at the University of Plymouth. The Centre is one of the HEFCE-funded Centres for Excellence in Teaching Learning (CETL). This CETL in Education for Sustainable Development aims to promote institutional change in the areas of curriculum, campus, community and culture.

**Chris Hails** is an ecologist trained in Scotland. He taught at the University of Malaya in the late 1970s and was an environmental advisor to the Singapore government in the 1980s. He joined WWF International in Switzerland in 1988 first as Director of the Asia Programme, later as Global Programme Director.

**David Harmon** is the executive director of the George Wright Society, an international association of researchers, managers, and other professionals who work on behalf of protected areas. He is also vice chair for North America of IUCN's World Commission on Protected Areas. He maintains an active research interest in the relationship between biological and cultural diversity, having co-founded the NGO Terralingua, which is devoted to that subject. Harmon is the author of *In Light of Our Differences: How Diversity in Nature and Culture Makes Us Human* (Smithsonian Institution Press, 2002) and co-edited (with Francis P. McManamon and Dwight T. Pitcaithley) *The Antiquities Act: A Century of American Archaeology, Historic Preservation, and Nature Conservation* (University of Arizona Press, 2006).

**Peter Howard** founded the *International Journal of Heritage Studies* and edited it for 12 volumes. He studied geography, and later taught landscape and heritage issues at Exeter College of Art. He is now retired, but attached to the Landscape Research Group and Bournemouth University—one of the few to include natural and cultural heritage within one course. He is author of *Heritage: Management, Interpretation, Identity* (Continuum, 2003).

**James Kushlan** is an ecologist, ornithologist, and conservationist. He has published widely in the areas of wetland ecology and aquatic ornithology, including most recently the books *Heron Conservation* (Academic Press) and *The Herons* (Oxford), and was the lead author of *Waterbird Conservation for the Americas*. He has served as professor and chair of biology at the University of Mississippi, Director Patuxent Wildlife Research Center, and Research Associate at the Smithsonian Institution. He has also been President of the American Ornithologists' Union and the Waterbird Society, Chair of the Heron Specialist Group, and founding Chair of the North American Waterbird Conservation Plan. He lives in Key Biscayne, Florida, and Annapolis, Maryland, USA.

**Jean-Claude Lefeuvre** has been professor at the Université de Rennes and, since 1979, at the Muséum National d'Histoire Naturelle in Paris where he created the Chair of Applied Ecology in the 'Laboratoire d'Evolution des Systèmes Naturels et Modifiés'. At the same time he has directed CNRS and INRA research teams, and has been responsible for many research programmes on the functioning of salt marshes on the west coast of Europe. He has presided over the Fédération Française des Sociétés de Protection de la Nature, and was vice-president of WWF France and president of the Scientific Council. He is currently president of the Institut Français de la Biodiversité.

**Thymio Papayannis**, an architect and planner, founded a large consultancy in 1958 responsible for the design of major projects in Greece and abroad. In the early 1990s he contributed to the establishment of WWF Greece, the Society for the Protection of Prespa and the MedWet Initiative (for many years president of the first two and co-ordinator of the third). In 2003 he founded Med-INA (Mediterranean Institute for Nature and Anthropos) and is its director. He has written more than 200 articles and five books on environmental and planning issues, and especially wetlands and the nature–culture interface. Among other activities, he is at present co-ordinator for the National Spatial Plan of Greece and a Trustee of WWF International.

**John Scott** is Programme Officer for Traditional Knowledge at the Convention on Biological Diversity, and had an academic and UN career before his present post. His interests are in bringing indigenous heritage issues to bear on the conservation of natural heritage.

**John Sheail** is a Research Fellow of the Centre for Ecology and Hydrology (Natural Environment Research Council) in the UK. Prior to his retirement, he was Deputy Head of its Monks Wood site. His 200 or so publications have ranged over a wide range of themes in environmental history and, more specifically, those of historical ecology, the emergence of the environmental sciences, and the interface between environmental research and public policy making.

**Eileen M. Smith-Cavros** is an environmental sociologist who has published articles and performed research on the topics of race and ethnicity, spirituality, and the relationships between people and their natural environments. She was the founding director of the Zion Canyon Field Institute in Zion National Park. Her book *Pioneer Voices of Zion Canyon*, concerning Mormon pioneers, place, and natural resource use, was published in autumn 2006. In south Florida, her research examined the atala butterfly, including the roles people have played in the destruction and preservation of the species. Currently, Eileen is an Assistant Professor at Nova Southeastern University in Florida.

# Editorial: Nature as Heritage

## Thymio Papayannis & Peter Howard

### Theoretical Considerations: The Intrinsic Value of Nature

After the turn of the 21st century and during the first half of the current decade, a number of prominent international organisations[1] mobilised thousands of scientists throughout the world to produce the Millennium Ecosystems Assessment (MEA), which documented the value of nature for human beings and the degradation to which it continues to be subjected. One by-product of the MEA is the concept of 'ecosystem services'—services rendered by nature to humans of significance for combating poverty, encouraging development and improving the quality of life.

Thus, the pendulum seems to be swinging once more from the direct appreciation of nature by scientists, naturalists and explorers—during the first part of the 20th century—to a utilitarian view, often expressed in financial terms, through the 'valuation' of natural assets.[2] This approach is adopted even by strong conservationists who believe that in this way their case for nature protection can be more easily accepted by political forces and the business sector.[3]

And yet there are those who continue to maintain that nature has intrinsic value and that it should be protected for its own sake. This consideration is often based on spiritual or metaphysical beliefs, but it also results from moral considerations and the responsibilities of human beings towards the natural world.[4] An elegant intellectual construct has been formulated through the Gaia theory, which attributes to our planet as a whole the characteristics usually associated with life forms, including self-defence.[5] The message is strengthened by a new sense of urgency, resulting from a growing awareness of the dire impacts of climate change, which are already highly visible.[6] Thus, the focus may shift rapidly from long-term efforts to maintain and optimise the services to humanity provided by ecosystems to urgent actions in order to avert major climatic catastrophes.

In parallel, cultural scholars, especially those involved with cultural landscapes, are becoming aware that nature is an inextricable part of culture, thus reinforcing the non-utilitarian view. This new realisation takes concrete form in the designations of Cultural Landscapes within the UNESCO World Heritage Convention, as well as in the framework of the European Landscape Convention,[7] which considers landscapes as the composite result of both natural processes and human activities. The Convention also clearly includes within the definition of 'landscape' that it is 'as perceived by humans', thus strengthening the utilitarian position.

The growth of this utilitarian position, and similar downgrading of intrinsic values, can be very clearly mirrored within the community of cultural heritage. The intrinsic values once thought to reside almost within the stonework of historic buildings are now more often regarded as cultural values imposed upon the building by scholars, and that others from other disciplines or other perspectives impose quite different sets of values on the same piece of heritage. Indeed, the very concept of 'authenticity', once so routine within cultural heritage, has now been shown to be anything but immutable or intrinsic. Different disciplines use different authenticities.[8] But the value of heritage to communities is now seen as its greatest asset, and its greatest claim to funding. Not only is the conservation of the cultural heritage seen as fundamental to the development of the tourist industry, hence the enormous demand for World Heritage status, but it has been used successfully to regenerate cities and regions irrespective of their tourist potential. The French ecomuseums, for example at Le Creusot, are examples of this, as well as some major cities such as Glasgow.

## Nature and Human Beings

The intimate and complex relationship of human beings with nature characterise almost the entire era of the presence of *Homo sapiens* on this planet. In fact, for most of this lengthy period, humans were part of nature and only relatively recently started modifying it to serve their needs better for food and security. In our Western industrialised and 'wired' civilisation, the relationship with nature is being weakened under the dual pressures of urbanisation and information technology. Throughout the world, but especially in technologically developed or developing countries, the population is moving to urban centres and the resulting urbanisation 'consumes' rural and natural landscapes at a rapid rate. This is compounded by the unsustainable exploitation of natural resources (such as timber, fish and hydrocarbons). In addition, information technology encourages indirect contact with nature through the establishment of virtual reality.

Yet interest in nature remains lively, though it may be limited perhaps to particular peoples (many of them indigenous) and minorities. Paradoxically, in the more affluent countries, a considerable number of people are looking for better living conditions outside urban centres, and preferably in well-preserved rural areas and in proximity to protected natural sites or cultural landscapes. Hobbies related to nature (such as gardening, bird watching, hiking, mountain climbing, skiing, scuba diving and kayaking) remain highly popular. In fact, the growing participation in such recreation activities is degrading natural assets, even in protected areas. Many major tourist centres trade primarily on their natural ecosystems, including wildlife. A simple content analysis of any world tourist brochure will illustrate the enormous part played by animal life in attracting visitors to many parts of the world, notably Africa and the tropical seas. In part at least, nature is being loved to death.

Thus, the future of the relationship between humans and nature remains unclear, especially when viewed in the framework of the predicted (and highly probable) large-scale destruction during this century resulting from global climate change. The sheer speed of the predicted change (and indeed the observed change) is likely to ensure that losses of species and habitats greatly outnumber gains, besides its impact on humanity.

## Nature as Heritage

Viewing nature as heritage has a double impact. On the one hand, it associates nature with human beings, as 'heritage' and 'inheritance', let alone 'patrimony', are highly anthropic terms (at least in this context). On the other, it implies a sense of responsibility of the natural wealth that has been received from our ancestors and the natural dividends that we shall leave to our descendants—unless we manage to 'bankrupt' the natural world.

In this sense, considering nature as heritage does contribute to a better realisation of its value for the present and future generations. It may lead to greater social and individual responsibility and thus contribute to nature conservation efforts. The notion of heritage is strong in relation to culture and it has been the basis for the protection and enhancement of the cultural inheritance. Very large sums of money have been poured into 'heritage' by governments at all levels and particularly by nation-states attempting to bolster their legitimacy. But even these sums pale into insignificance beside the time and money expended by private citizens on their own property, and attempting to pass on that property and their values to their children. Similarly, the understanding of the value of natural heritage (such as high biodiversity, functional terrestrial and marine ecosystems, clean air, ground and water, intact landscapes, stable climatic conditions, reversed desertification and erosion) may be the most valuable inheritance that we should leave to our grandchildren. If human beings can accept the need to regard nature as their heritage, as private individuals as well as members of a polity, the rewards might be very great indeed.

## Places

Natural heritage can be approached in many different ways. Some of them are treated by the papers included in this special issue, which aim to be representative but by no means exhaustive. Their main purpose is to encourage debate and lead to a continuing exchange of views between the natural and the cultural sectors.

A first and obvious approach is the geographic one. While culture is primarily society specific, nature is geographically specific as a result of climate, geology and geomorphology and many other factors.

*Regions and Countries*

The two first papers refer to specific geographical areas to make their points. A first paper by Giorgos Catsodorakis[9] deals with the natural heritage of the Mediterranean Basin and more broadly in Europe. John Sheail[10] focuses on the UK. Their views on the nature/culture relationship are spherical, but far from identical. They both provide conclusions that are of pertinence to other countries and other continents. Thus Catsodorakis, writing on 'The Conservation of Natural and Cultural Heritage in Europe and the Mediterranean: A Gordian Knot?' argues that, in order to discuss a more effective conservation of the European natural heritage, three issues must be addressed. To start with, the terms 'conservation', 'natural heritage' and 'cultural heritage' must be more carefully defined, as people perceive them differently, and in our globalised modern world this situation creates confusion, especially during the decision-making process. Indeed, the need for a common language is a constant theme throughout this special issue, juxtaposed with the expressed need to conserve as many languages as possible. Secondly, it is not possible to approach the subject of natural heritage, anywhere in the continent, without a direct reference to the socio-cultural processes that have contributed decisively to its creation. It was for this reason that the European Landscape Convention has not included the word 'Cultural' in its Convention title, assuming that all European landscapes are indeed cultural, at least in part. Finally, the tools which the EU uses in order to define and conserve the natural heritage are not appropriate for this major task. It is thus argued that the intangible cultural heritage that shaped the European natural wealth is not systematically recorded and maintained. In conclusion, it is proposed that the safest and most effective way to conserve the European natural heritage is through the support of its traditional agricultural and forestry sectors. This same need to make contact with 'traditional knowledge' has been the subject of a major conference in Florence, within the forestry community.[11]

Sheail's paper, on the other hand, '"One and the Same *Historic* Landscape": A Physical/Cultural Perspective', maintains that ecologists have been striving for years to define and bring greater rigour to the term 'naturalness'. With the assistance of historical ecology, an attempt is made to advance the study of landscapes and contribute to their conservation. The paper guides us through the history of building the institutional framework for the protection and the custodianship of the natural heritage in the UK. This history is far from unique, and has been influenced by, and influenced, other countries. This endeavour is greatly supported by the knowledge of the ecological history of each plant and animal community. Such information can be collected from literary, archival and oral memory records, as well as from evidence that has remained and is preserved within the landscape itself. One quirk of the English system (not shared even by other parts of the UK) has been the clear differentiation between landscapes conserved for cultural reasons, including the National Parks, and places conserved for natural heritage, such as Nature Reserves. There have been ongoing attempts to erase this odd distinction, but it has meant that the larger protected areas, the National Parks, have always accepted the need for public access and the management of farming and forestry.

*Sites*

Looking at specific sites, Jim Kushlan[12] writes about the Everglades in Florida, while Jean-Claude Lefeuvre[13] examines the case of Mont St Michel on the Atlantic coast of France. The interest of these two case studies is that the first illustrates how one of the greatest wetlands of the world is as much cultural as natural; while it is documented that the second, a famous cultural site, has been developed as much by nature as by human beings.

Thus Kushlan and co-author Eileen M. Smith-Cavros, writing on 'Human Heritage and Natural Heritage in the Everglades', show that, since the beginning of human presence in the Everglades wetland, human inhabitants attempted constantly to manipulate their environment for better chances of survival. From the time when they first settled in the marsh complex (approximately 10,000 years ago) until the resettlement of south Florida at the end of the 19th century after its virtual abandonment for 200 years, local inhabitants showed a remarkable respect for the environment, exploiting nature to the extent that their abilities and knowledge allowed. This all changed in the late 1880s, when technology enabled humans to do what their ancient predecessors could not—alter their surroundings dramatically. Excluding the Native American part of the population, modern settlers of the Everglades generally regard themselves as temporary residents, from holiday-makers to refugees, an attitude that supports the exploitative character of their approach to the natural environment. The situation has been further exacerbated by the fact that most of the new settlers originated from countries where nature was perceived as infinite or having little value. Now humans appear dominant, but it remains to be seen whether their reign will be long lasting.

In 'Natural World Heritage—A New Approach to Integrate Research and Management', Lefeuvre argues that nature, in its totality, has been influenced decisively by humankind. When climatic conditions became favourable (ca 9,000 years ago), humans took a great revolutionary step in abandoning their nomadic way of life and adopting a sedentary lifestyle, with the dawn of the Neolithic Age and the establishment of agriculture. Little by little humans began to alter the natural environment and adapted it to serve their ever-growing needs. Its inevitable degradation necessitated, in relatively recent years, the establishment of measures for its protection. These general considerations are documented by the case of Mont St Michel and its bay, the first French UNESCO World Heritage Site, its course through the centuries and especially the changes it has undergone during the last 100 years. Despite the twice-daily incursion of the sea, the detailed topography of the entire bay, its wildlife and entire ecology are all determined by successive cultural uses. The bay is almost as much a cultural artefact as the abbey at its centre. More positively it constitutes an example of how serious and effective management plans can contribute to the reversal of environmental deterioration.

## Managing Natural Heritage

*Conservation*

For all the reasons mentioned above, conserving the natural heritage has been a major concern of individual scientists, laymen and intellectuals for more than a century.

Starting from the individual concerns of enlightened individuals, it gradually became adopted by non-governmental organisations and finally became incorporated in national and international policies.

A critical view of the changing attitudes and practices in the conservation of the natural heritage, with reference to characteristic organisations active in this sector, is provided by Chris Hails's paper[14] on 'The Evolution of Approaches to Conserving the World's Natural Heritage: The Experiences of WWF'. WWF was founded in 1961 by a small group of mostly British naturalists and conservationists, with the intention of conserving 'the world's fauna, flora, forests, landscape, water, soils and other natural resources'. In the 1970s public awareness grew, but so did the realisation that economic development had a severe impact on the planet's resources. In 1980, WWF, along with IUCN and UNEP, produced the World Conservation Strategy and after the Rio Summit in 1992 it became evident that a sound environment was the essential prerequisite for all human development and welfare. As a consequence, it was considered necessary to make an effort to alter the role of business at the global scale, as the impact of unsustainable business practices constituted a heavy burden upon the destitute of the world and the natural environment.

According to Hails, WWF and other major organisations have developed strategies that provide a geographical focus and on-the-ground experience that facilitate the delivery of environmental solutions. Its activities are targeted mainly at the national level and have a local scope, increasing the prospects of success. It is believed that where political will is lacking, democratic processes and creativity can contribute to the re-establishment of human society's interdependence with nature, a state that past generations enjoyed for thousands of years, before the term 'sustainability' was invented. WWF also stands as an exemplar of the success of non-governmental organisations at international, national and local levels in their efforts to mobilise public opinion and private wealth.

*Trans-sectoral Dialogue and Education*

Very few scholars have been able to bridge the divide between ecology and conservation on the one hand and cultural disciplines on the other. The close links between culture and nature that emerge from the two papers by David Harmon[15] and Alan Dyer[16] in this issue require as a minimum further exchanges between what can so easily be seen as two sides. However, if sides there are, then there are many more than two, as there are as many significant disputes between scientific disciplines as between cultural disciplines, and certainly between the academic expert world and other stakeholders—as the need for education so clearly demonstrates. There is strong need for instituting and developing dialogue between those responsible for the study and conservation of the natural heritage and those involved with the cultural heritage, with the aim of increasing understanding and obtaining synergy.

Harmon argues in 'A Bridge over the Chasm: Finding Ways to Achieve Integrated Natural and Cultural Heritage Conservation' that for many years conservationists and social scientists have been sitting on opposite sides, as far as integrated natural and cultural heritage conservation is concerned. This has resulted in a barren confrontation that has done little to promote the objectives of protected area management. The situation is an outcome of different orientation, which stems from diverse training and educational background and causes distrust on both sides. Although common efforts appear difficult to achieve, the development of biocultural diversity research aspires to bridge these differences. Scientists from both fields are required to work closely in order to establish a common ethical base and to support interdisciplinary efforts. Successful results of this endeavour will lead to more effective integrated conservation of the natural and cultural heritage. Such disputes may sometimes seem to be largely theoretical, but there is scarcely any conservation project that at some point does not have to weigh carefully the priorities of the conservation of some natural habitat and that of some cultural artefact. Surprisingly often it is possible to do both, once the problem is recognised.

Dyer, on the other hand, in 'Inspiration, Enchantment and a Sense of Wonder … Can a New Paradigm in Education Bring Nature and Culture Together Again?' maintains that if we are to change our world for the better then environmental education, addressed both to young children and older students, will play a substantial role. Educational methods must change from a transmissive to a transformative mode, so that climate change, habitat loss, degradation of cultural heritage, threats to biodiversity and ecological stability can be addressed effectively. The paper further addresses issues such as how to define sustainability in a world in which the human population is growing immensely or how to achieve an appropriate holistic training that includes both the sciences and the arts. Although outstandingly effective educational programmes are put into practice worldwide, their use must be encouraged and the good practice must be shared, but the pressures to reduce or even eliminate learning in the field, for both financial and legal reasons, seem to grow continually.

*Multilateral Responsibilities*

At the interface of nature and culture, multilateral agreements are playing an important role in improving understanding and collaboration. This may be considered obvious in the case of UNESCO, which has a mandate for both science and culture, as well as education. In addition, it has launched the World Heritage Convention with sites designated either for their cultural or their natural importance.[17] Mixed sites have been recognised from the beginning, and to these has now been added the category of Cultural Landscapes, which inevitably involve both nature and culture. Its example, however, has been followed by purely 'physical' multilateral organisations, such as the Conventions on Biological Diversity, on Wetlands and on Desertification, which have begun incorporating cultural aspects in their work, as documented by Peter Bridgewater, Salvatore Arico and John Scott.[18]

In their text on 'Biological Diversity and Cultural Diversity: The Heritage of Nature and Culture through the Looking Glass of Multilateral Agreements' they argue that it has been only relatively recently that the role and importance of cultural heritage and diversity have been understood by the majority of the scientific world. Even now they are often viewed in contrast to the natural heritage and to biological diversity. Instead, these two aspects reflect the two sides of the same coin and should be examined in this light, as the global heritage of humankind. However, Multilateral Environmental Agreements (MEAs) often fail to take into consideration the fact that the two diversities are mutually self-supporting. Lately, though, this trend has begun to change and many MEAs support the interdependence of biodiversity and of cultural diversity. Defining the research agenda on the interlinkages between natural and cultural heritage constitutes a major issue, and the international organisations should prioritise specific programme activities in their work planning in order to achieve better results.

*The Transcendental Dimension*

Beyond natural and the more normally considered cultural aspects, spiritual considerations must also be taken into account, as they may play a significant role in the protection of human heritage. Religion has been a major component of any culture, and this becomes of particular significance as, at the turn of the millennium, not only is there a considerable upturn in several major faiths, but many are showing an increased interest and concern about environmental issues, and in particular the natural heritage, viewed by them as part of the divine Creation and, therefore, sacred. This is manifested through concrete initiatives, aimed at sensitising world public opinion and mobilising the corresponding religious structures. Such environmental considerations, related to traditional religious disciplines and beliefs, concern not only the Christian faith and its main branches (Catholicism, Orthodoxy and Protestantism) but also Buddhism, Islam, Hinduism and Judaism, as well as a multitude of other faiths, such as the well-known concern of the Jains for all life. In addition, increasing co-operation among the major faiths is often focused on environmental matters with increasing joint initiatives.

A characteristic example is HAH The Ecumenical Patriarch Bartholomew I of the Orthodox Christian faith.[19] In his text here on 'The Gift of Environment: Divine Response and Human Responsibility' he argues convincingly that if we want to deal with the problems of our environment effectively then we must change the way we envisage the world, for the crisis we face is not principally ecological. He further maintains the religious view that the world we live in and enjoy is a gift from above. It was bestowed upon our generation, which should appreciate it and thank the Creator for His generosity. Human beings need to overcome their innate greed and realise that the environment is not ours to abuse and waste, but instead it is the home of every form of life created by God. According to Patriarch Bartholomew, we should preserve it as best as possible and bequeath it to our children by adopting an 'ascetic ethos', manifesting self-restraint and love for the entire Creation.

## Conclusions: The Way Forward

Although diverse in focus and approach, all the texts presented point out the need to integrate the approach to the cultural and natural heritage. In fact, this has been the major objective of this publication, as up to now the publishing world has been as divided as the academics. The boundaries between publishers' lists are as impenetrable as those between university faculties. This one special issue should be considered, however, only one of the first steps. Many others are necessary.

Developing a common language is one of the key priorities, so that the two sides can communicate in a more effective way. This should emerge from joint research, and needs common vehicles for publication, of which the *International Journal of Heritage Studies* must certainly be one. The research methodologies may well be different. Culture is only occasionally susceptible to the kinds of reproducible experiment and quantitative analysis that are routine in the natural sciences, and it is easy for scientists to denigrate the qualitative methods that so often lead to breakthroughs in the humanities and social sciences. And far too many cultural specialists regard the techniques of the scientist as pedestrian and even outdated. We need to comprehend better how human activities, which constitute culture in its broadest definition, transform nature in both positive and negative ways. We also need to understand the influence of nature on culture in different periods and in diverse societies. Culture, of course, is natural, as nature is cultural.

Education at all levels will play a key role in bringing the two sides together, not just in the school but in the academic workshop. New disciplines and approaches, which look at heritage in an integrated manner, may be required here, and may make prioritising between various fields of heritage much simpler. So far, Heritage Studies has emerged mainly from the cultural faculties, though from a great variety of disciplines. Most courses do not attempt to bridge the gap, but if students from these courses cannot work equally with all the fields of heritage conservation then there is little hope for those from more traditional disciplines.

Finally, we need to preach convincingly to physical scientists and conservationists about the importance, pertinence and the delights of culture and to cultural scholars about the magnificence and omnipresence of nature.

## Notes

[1]   Such as the United Nations Environment Programme (UNEP), WWF International, IUCN, BirdLife International, Fauna and Flora International, The Nature Conservancy, Wetlands International and the Wildlife Conservation Society.

[2]   Barbier et al., *Economic Valuation of Wetlands*.

[3]   As demonstrated by the broad and powerful impact of the Stern report on climate change.

[4]   A characteristic example is the work currently carried out by WWF UK through a project on the values of nature led by Tom Crompton.

[5]   Margulis, *The Symbiotic Planet*.

[6]   Brown, *Global Warning*; Lovelock, *The Revenge of Gaia*.

[7]   A key factor in individual and social well-being and people's quality of life, the landscape contributes to human development and serves to strengthen European identity. The aims of this Convention, in the framework of the Council of Europe, are to promote landscape protection, management and planning, and to organise European co-operation on landscape issues.

[8]   See Ashworth and Howard, *European Heritage Planning & Management*.

[9]   Biologist and conservationist, scientific advisor of WWF Greece.

[10] Centre for Ecology and Hydrology, Natural Environment Research Council, UK.

[11] International Conference on Cultural Heritage and Sustainable Forest Management: The Role of Traditional Knowledge, Florence, Italy, 8–11 June 2006 (http://www.forestland-scape.unifi.it/).

[12] Professor, University of Florida, USA.

[13] Professor, University of Montpelier and CNRS, France.

[14] For many years Chief Scientist of WWF International, Hails is now Director of Network Relations of the same organisation.

[15] Executive Director, the George Wright Society, USA.

[16] Principal Lecturer in Environmental Education at the University of Plymouth, UK, and co-author of *Let Your Children Go Back to Nature*.

[17] The Convention Concerning the Protection of the World Cultural and Natural Heritage is an international agreement that was adopted by the General Conference of UNESCO in 1972. It is based on the premise that certain places on Earth are of outstanding universal value and should therefore form part of the common heritage of mankind.

[18] (a) Secretary General of the Convention on Wetlands (until July 2007), (b) Programme Specialist, Biodiversity Division of Ecological Sciences, UNESCO, and (c) Commonwealth Scientific and Industrial Research Organisation, CSIRO, Australia, and Convention on Biological Diversity, respectively.

[19] Also known as the 'Green Patriarch' because of his many ecological initiatives, such as the ecological symposia on the Aegean, the Black Sea, the Danube River, the Adriatic and Baltic Seas, and in 2006 the Amazon.

## References

Ashworth, G. and P. Howard. *European Heritage Planning & Management.* Exeter: Intellect, 1999.

Barbier, E. B., M. Acreman and D. Kowler. *Economic Valuation of Wetlands: A Guide for Policy Makers and Planners.* Gland: Ramsar, 1997.

Brown, P. *Global Warning: The Last Chance for Change.* London: A&C Black, 2006.

Lovelock, J. *The Revenge of Gaia.* London: Allen Lane, 2006.

Margulis, L. *The Symbiotic Planet.* London: Phoenix Press, 1998.

# The Conservation of Natural and Cultural Heritage in Europe and the Mediterranean: A Gordian Knot?

Giorgos Catsadorakis

### Defining 'Conservation', 'Natural Heritage' and 'Cultural Heritage'

Everyone, from laymen and activists to politicians and scientists, seems intuitively to understand and agree with a statement such as: 'The natural and cultural heritage (of a site) must be conserved'. As obvious and routine as this may be, if one wishes to go a step forward and implement specific actions, major problems emerge in adequately defining—particularly operationally—all three main terms: 'conservation', 'natural heritage' and 'cultural heritage', although they all sound familiar, are firmly established and widely used. It is beyond the scope of this paper to explore the content of these terms or to review the vast relevant literature. In order to serve my aim, however, some important points must be mentioned.

(1) We cannot exactly define what conservation means in practical terms because the subjects of conservation are continually changing and because deciding what exactly we want to conserve is not a purely scientific matter, so it can never have unequivocal, objective answers.[1] If we want to conserve a 'natural' or 'pristine' situation, the answer is that there is no rational scientific basis on which to define 'natural'[2] and since ecosystems with or without human interference change continually, there is no pristine situation. The science of conservation biology may inform the whole process of conservation activities, but decisions on what to conserve are arbitrary and mainly a matter of science, values, expediencies and politics. The best approximation to a definition of conservation is 'creating conditions that allow ecosystems to change, with the least species loss and the least damage to ecosystem processes'.[3] This is not very helpful, however, in decision making on what, how much, when and for how long.[4] Generally, conservation entails both site protection and site management and should be conceived along four main axes: genes, species, habitats, landscapes. Unfortunately so far, only species and habitats have received sufficient attention from the international community.

(2) Natural heritage must necessarily relate to a specific place and time period and should at least include: specific sites, types of landscapes, species (and genes in the form of sub-species, races, varieties) and habitats. As ecosystems change continuously and as the human impact upon them also varies, natural heritage has a continuously changing content too. Since it is defined always in relation to the past (and present), the natural heritage of a place is also contingent upon the extent of our knowledge of past situations. For example, the natural heritage of the classic land of Attica, Greece, today is not like that of 500 years ago and certainly not like that of 5,000 years ago. The knowledge, however, we possess about the nature of Attica at that time is meagre. So the benchmark situation to be used to represent the natural heritage of Attica remains unresolved.

(3) Cultural heritage is composed of tangible and intangible elements.[5] The latter include language, legends, myths, norms, perceptions, practices, habits, customs, diets, methods, etc. Problems similar to those in (2) above also pertain to the notion of cultural heritage, which is itself a continuously changing concept, with much quicker rates of change than natural ecosystems, sometimes occurring in leaps.

Thus, to move forward in a concrete and useful manner we must inevitably compromise and accept a large amount of vagueness. Because the necessities to preserve our natural and cultural heritage are real and pressing, we must manage to achieve our goals despite this paradox.

**The Spatial Dimension: The Mediterranean in Europe**

The Mediterranean basin is probably the most difficult region to overview, because of its geological, biological and geopolitical diversity.[6] There are not always enough statistics for the northern part of the basin, which is a part of Europe, but the main features of the Mediterranean are described here, referring mostly to its northern coast and whenever necessary to the entire basin. Generally, whatever refers to Europe also holds true for the Mediterranean. In general terms, what sets the Mediterranean apart, or simply at the other end of the continuum, is its much higher biodiversity, a greater variety of landscapes and management practices,[7] and a resulting higher cultural diversity; above all, it shows an intricate patchwork of habitats with a far finer grain (smaller average patch size) than the rest of Europe. This results from its more intense relief, its more varied geological history, the mosaic of local climates, the relatively diverse effect of the glaciations and, of course, the fact that it is the cradle of very ancient civilisations, it is more or less densely populated and it has been heavily influenced by humans for at least 10,000 years.

In his worldwide analysis of biodiversity, Myers[8] rates the Mediterranean basin as one of the world's 18 hotspots. However, the basin is far too extensive and heterogeneous to be treated as a single hotspot area.[9] The Mediterranean is one of the world's major centres for plant diversity, where 10% of the world's higher plants can be found in an area representing only 1.6% of the Earth's surface.[10] Also, as many as 366 bird species currently breed in the 3 million $km^2$ of the Mediterranean basin, compared to the 490 breeding in the 10 million $km^2$ of the whole of Europe.[11] It also includes 45% of the bovid varieties and 55% of the goat varieties of Europe and the Middle East.[12]

On the other hand there is growing local demand for tertiary activities, especially near the coast, from promoters, speculators and entrepreneurs of all sorts. So, a dichotomy exists throughout the basin; far from the coast there is abandonment and woodland encroachment while along the coast all contact with the ecological and cultural past of the Mediterranean is being lost.[13] Yet the Mediterranean is still the number one tourist destination in the world, currently drawing more than 30% of annual tourist trade mainly from countries of northern Europe.[14] The Mediterranean also includes most of the areas of priority for conservation in Europe, since these are subject to rapid loss under current policies.[15] Hereinafter, by 'Mediterranean' I refer only to northern Mediterranean countries and especially those belonging to the EU.

**In Europe, and Particularly in the Mediterranean, Natural Heritage Cannot be Conceived Separately from Cultural Heritage**

> After 10 000 years or more of 'co- habitation' most Mediterranean ecosystems are so inextricably linked to human interventions that the future of biological diversity cannot be disconnected from that of human affairs.[16]

There are hardly any places on earth that do not bear the traces of human influence, and the few cases have certainly been the exception for hundreds of years. In Europe every place, with the possible local exceptions of steep cliffs and the highest mountain tops, is culturally modified.[17] It is fully documented and widely accepted that most of Europe consists of cultural landscapes. Farming accounts for more than 60% of the land surface of the EU, though less than 10% in Fennoscandia where forests predominate.[18] In the Mediterranean, half the land is occupied by agriculture and the rest by forests, mattorals and range lands[19] while the statistics are certainly confusing in cases where both farmland and shrubland are used for grazing. Agriculture and livestock rearing is also the major land use inside most of the protected areas in Europe.[20] This situation is similar but even more pronounced in Mediterranean countries as a result of a longer human presence. In fact some scientists have talked about a complex 'co-evolution' that has shaped the interactions between Mediterranean ecosystems and humans.[21]

Throughout Europe many people still suffer from the misconception that most areas important for nature conservation are wilderness areas. In fact, much of what is of value throughout Europe is not wilderness but farmland.[22] The majority of the 20,000 protected areas presently existing in Europe are also dominated by cultural landscapes.[23]

It is now well understood that the present situation of the natural environment of Europe is not a product of eons of natural processes alone but also of centuries of farming, livestock rearing, forestry and other kinds of management by humans. Figure 1 makes clear that, whatever the exact meaning of the terms 'conservation' or 'natural heritage', in order to conserve the natural heritage of Europe, either within or outside protected areas, we have to manage in an appropriate way the anthropogenic factors (i.e. mainly farming and forestry) that contribute to the creation of this heritage, since our potential to affect actively the natural factors is considered negligible and will not be discussed any further. Additionally, it is emphasised that there is no practical alternative, other than through farming, livestock rearing and forestry to sustain those landscapes, habitats and wildlife communities that are valued as our natural heritage.[24] The human element cannot be divorced from nature conservation issues in the European countryside.[25]

Of course, the proper management of human activities may entail that in certain, generally few, cases the best solution for the protection of particular natural heritage elements would be totally to exclude human presence, in which case we are talking about site 'protection'. Especially in the densely-populated Mediterranean, absolute exclusion of humans from very small areas will always be important to avert over-disturbance, but this cannot be the principal conservation strategy.

Over millennia, farming and livestock husbandry have developed practices and methods which have produced a remarkable structural and habitat diversity at landscape and regional levels. This includes natural hedges, terracing, grazing patterns, burning, transhumance, crop rotations, cultivars, varieties, livestock breeds, crop mixtures, mixed grazing–cultivation systems, ley patterns, forest openings, irrigation systems, harvest timing, transport and storing of crops and fodders, herd composition and movements, predator control, guarding against predators and pests, natural fertilisation, crop protection, provisions for wildlife, risk assessment and control, disposal of corpses, temporary animal enclosures, feeding patterns, livestock diversity, ploughing and soil protection techniques, traditional architecture, quality of materials,

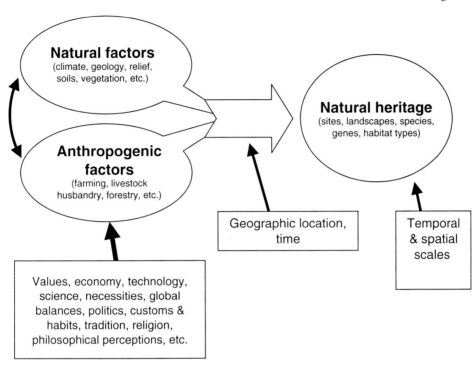

**Figure 1** Schematic representation of the relationship between natural and anthropogenic factors for the shaping of the natural heritage. Arrows denote influence.

aesthetic value of infrastructures, and so on. The temporal and spatial combinations of all these are also included. These human interventions provide disturbance to ecosystems and it is now widely accepted that small-scale disturbances taking place frequently result in higher species diversity than heavy disturbances or the absence of disturbance. Whether or not human intervention is totally excluded, ecosystems as living systems would go on changing because natural factors continue to act in any case.[26] Hence, one way or another, conservation of biodiversity or of landscapes or both entails a certain level of management.

Given all the above there are two important problems:

(1) As we have seen, management of ecosystems is necessary. In most cases rehabilitation and restoration will also be necessary since ecosystems change continually, but what conservation science alone will never be able to answer is what the goals of this management effort will be. What past biotic community that existed in a given area should be selected as the target for restoration efforts or as a benchmark situation? The answer to this question is by no means obvious. There is confusion even among conservationists about what exactly conservation means in a cultural landscape. As Rackham put it '... Ecological science as a whole has still not understood what conservation means in a cultural landscape'.[27]

(2) Even if it were possible to define such goals, it is commonplace that the decisions about what these patterns and mixes of farming, forestry and other human activities would be are not governed by the will to shape or maintain the natural heritage but rather from combinations of other much more influential factors, at various spatio-temporal scales such as values, economy, markets, technology, science, necessities, global balances, politics, customs and tradition, religion, stochastic events, philosophical perceptions, and so on. Collectively, all of these factors or parts of them constitute what can be called 'cultural heritage'; as we have seen, this is another very ill-defined concept, especially in its intangible constituents.

To give an example, in the 1970s the lesser kestrel (*Falco naumanni*) was a common, locally very abundant and widespread bird in Greece. It is a species that nests especially on roofs of old-type, derelict farmhouses and barns and which needs large steppe-like grasslands (or their human replacement—extensively cultivated areas with cereals) over which to hunt for insects, its main prey. Within three decades, with dramatic changes in farming practices and the built part of the countryside, it became a globally threatened species. At present it survives in some pockets of marginal lands within Thessaly, the most intensively cultivated area of Greece, and in some other areas which, because they are remote and isolated, still retain old-style farmland, such as the island of Lemnos. These remnant pockets of suitable habitats are, of course, under strong pressure for transformation and there is evidence that both agricultural intensification and marginal land abandonment have detrimental consequences for the survival of the lesser kestrel and other pseudo-steppe species.[28]

The lesser kestrel as a species is certainly a part of the natural heritage of Europe and should be conserved. The unanswered questions start when we have to take measures locally; where should it be conserved and what densities would be satisfactory? What will replace the loss of the old-style houses and their respective attributes for nesting? Science has so far failed to answer these questions and will never do so. Many scientists argue also that the high numbers and wide distribution of lesser kestrels recorded in the mid-1960s would never have been possible had it not been for the huge expansion of extensive cereal cultivation in recent centuries in Europe and the conditions of the built environment of settlements. So, they argue, it is now returning to a more 'normal' or 'natural' abundance and distribution. Thus, what period should be used as a benchmark for the normal or natural abundance of the lesser kestrel in order to be set as a goal for conservation measures in every location in Europe? And it must be a period for which we have data.

The conservation of natural heritage cannot be conceived without human management, and decisions about the implementation of farming, livestock rearing and forestry are taken within a framework created by a mix of present needs and the cultural heritage of European people. The situation is very complex, since not only do two ill-defined and only intuitively approached notions depend on another ill-defined notion but also the entire management system is not clearly visualised and described and there are no definite answers as to how aesthetics, the economy, the well-being of rural societies, food production and safety and (unclear) environmental goals will be synthesised in the right mixes and accommodated within the huge variety of local conditions of every single area in Europe in the right combinations, to ensure a satisfactory conservation of natural heritage. However, the need is there, real, and acknowledged by everyone.

## Conservation in Protected and Non-protected Areas

Until the beginning of the 20th century, production systems, landscapes and biodiversity were all accommodated within and coupled with certain social structures and societal organisations, technological tools and energy inputs. There was almost no concern and no need for the conservation of nature, or specific concern to create protected areas, etc. In other words, a certain culture and technology acted along with natural factors (climate, geographic position, stochastic events, and so on) on ecosystems to create certain kinds of landscapes and certain kinds and levels of biodiversity.

The notion of conservation of natural heritage per se emerged when centuries-old pre-industrial agro-sylvo-pastoral practices started to change, giving priority to the maximisation of efficiency in food and material production. This coincided with a dramatic change in technology, especially with the applications of internal combustion engines, increased energy and chemical inputs, and increased market globalisation. The natural heritage left by the pre-industrial rural Europe started to change precipitously. Species and habitats declined and vanished at variable rates from one place to another. Less-favoured or disadvantaged, mountain, very remote and island areas are still changing at a much slower pace.

Initially, the various countries following the American model of protecting the wilderness areas as part of the national heritage started to designate landscapes of national importance as protected areas, based on their natural beauty and various degrees of cultural importance. These included both some of the wildest places of the Continent, though, with only few exceptions, culturally modified, and a lot of clearly cultural landscapes. The criterion of biodiversity, or the conservation of species and populations, started to be increasingly incorporated into the criteria of protected area designations and as the dramatic loss of pre-industrial natural heritage was being appreciated, the shaping of a more European natural heritage context was promoted either through the Council of Europe or the EU or both (i.e. international conventions, EU directives, national laws, etc.). Today, there are about 20,000 protected areas in Europe, including the Natura 2000 network. These were either protected mainly as landscapes or to conserve species and habitats, in the later phase of designations. It is not clear whether all these include a satisfactory part of the European natural heritage but it is clear that almost the whole of Europe is a cultural landscape to a greater or lesser extent, and a great part of our natural heritage resides outside protected areas. The conservation of many animal species cannot be ensured only within protected areas, however large these may be. This applies especially to large and wide-ranging mammalian and avian carnivores as well as scavengers.

One thing is sure, conservation of natural heritage of Europe must be done both within and outside protected areas and in every case it must stem from a compromise between the well-being of farmers, the economy/market, consumer health and the conservation of species and habitats, taking into account the vast intra- and inter-area variation. As Perrings et al. have eloquently put it: 'Science has not yet addressed the trade-offs between food production, biodiversity conservation, ecosystem services and human well being in agri-cultural landscapes.'[29] Cahlin argues that 'farming systems and their ecology are too complex for anyone to know all the answers'.[30] Research, though, on pastoral farming systems and their interaction with the natural environment remains a priority.[31]

## A Critique of the Major Concepts for Natural Heritage Conservation in Europe

As is implicit in the previous section, it would make more sense to pursue the conservation of European and Mediterranean natural heritage through landscape conservation, the broadest of the four pylons of natural heritage in which the other three (genes, species, habitats) are embedded. However, site and landscape conservation is a very complex subject where aesthetics, land economy and management, and ecological well-being overlap.[32] How have these factors been considered at a European level?

Although in almost every European country there are official national inventories of the natural heritage, i.e. sites of national importance, such as National Parks, protected landscapes, and habitats and species and sub-species of national conservation concern, at the European level, to date and especially in the last years, nature conservation bodies have tended to concentrate on species and habitats (i.e. biodiversity) while landscapes have received limited attention only in some countries and mainly as past approaches.[33] Does this indirectly imply that while species and habitats can be a matter of Europe-wide concern, the conservation of landscapes is only of regional or national concern? And, if so, how can these be treated separately?

At present, the only concrete and unanimously accepted legal documents that have attempted to list part of the common European natural heritage are EU Directives 79/409 (Birds) and 92/43 (Habitats & species other than birds). These, in fact, list the rare, endemic and threatened species and habitats, present on the Continent, and their degree of rarity in the period 1976–2003, when the relevant inventories took place. These lists must not be understood as representing all the natural heritage of Europe and must not be allowed undue influence as:

(1) They fail to describe the entire biotic communities these species belong to, and the qualitative and quantitative composition of these communities as well as the landscape part of the natural heritage is totally missing.
(2) Unfortunately, the official environmental authorities of the EU direct funds only to the conservation of these species and habitats within the protected areas designated for that reason, i.e. the Special Areas of Conservation, which constitute the Natura 2000 network. It is already common knowledge, however, that the objectives of these two directives cannot be achieved by site management within protected areas alone[34] and, indeed, many species of valued birds and mammals

require large areas, often as intricate mosaics of farmland habitats, to meet their needs throughout the year. Limiting nature conservation to setting aside protected areas would thus be ineffective since they would either be too small or unsustainable in the long term.[35] Also the two directives do not provide adequate mechanisms to fulfil their objectives. These must be met in large part by the incorporation of conservation measures into the application of legislation dealing particularly with the agricultural and rural economy sectors.[36]

Every member state of the EU is responsible for adopting measures and activities to guarantee the Favourable Conservation Status of populations and areas, and the condition of species and habitats listed in these two directives. However, this favourable status must be defined qualitatively and quantitatively and, because there are intrinsic interdependencies of the constituents of ecosystems, the quantitative ratios between them must be real. Thus, there is always a need to refer to a benchmark situation where the overall ecological condition of an area or a site permitted the survival of, for example, one pair of golden eagles (*Aquila chryasetos*). There is no prescription for such Europe-wide decisions. These must be taken locally. The European natural heritage had been maintained in Europe, to the best of our knowledge, almost unchanged for some centuries until the mid-19th century, when the agro-sylvo-pastoral systems of the pre-industrial era started to change and the rates of change increased dramatically after the Second World War.[37]

So, we have not well understood how these complex systems had worked,[38] but we do have the lists of species and habitats we want to conserve. Since these can be conserved only within complex ecosystems they must be in certain relative proportions. We do not know these proportions and we do not know how we will achieve them today, because ecosystems continuously react to human and natural perturbations. Conservation science cannot give the answers; it can only inform the process. We know that, for centuries, old-fashioned farming and pastoral practices had been producing a situation that today we want to conserve or restore and at the same time support rural communities and produce safe foods. An answer would be to restore these old-fashioned methods and practices, but this is not possible because knowledge is being continually lost, the social context has changed irreversibly, the global situations are completely different and the technological tools and values are even more distant. However, we could possibly keep the essence, the unchanged nucleus of these practices that have not changed because they directly relate to the rhythms of nature. So, the question is: is there any officially recognised inventory of all these centuries-old practices, methods, proportions, timings, techniques and combinations across the Continent, together with local variations, that could possibly underpin such an effort? The answer is no.

## Conservation of Natural Heritage through Conservation of Cultural Heritage

We need to differentiate between the two ends of the continuum. The mainly lowland, industrialised farming systems are of very low biodiversity and lower perceivable heterogeneity, and so these are not discussed here. The low-intensity systems that are found mainly in protected areas, as well as in Less Favoured Areas, are of high value both for biodiversity and aesthetics.

First, it is essential to collect, describe and list these practices, accompanied, of course, by geographic ranges and the necessary groupings. They should be analysed to understand their implications for biodiversity and landscape and be officially recognised as part of our cultural European heritage. If this were done, it could be used for strengthening the weak concept of a common cultural farming heritage across Europe, but secondly, and more importantly, the conservation of the European natural heritage could be based upon the conservation of these practices and methods. The idea is simple, although the implementation would not be such—to channel subsidies for agriculture and livestock to the maximum possible use and implementation of a number of practices and methods that we would have already listed and declared as Europe's farming intangible cultural heritage. Nature would be left to do the rest, while conservation science would monitor the results and suggest rearrangements and fine tuning. Due to huge variations and local particularities, subsidies could be channelled not only to individuals but also to groups of farmers or to administrative areas.

The most recent reform of the CAP in 2003 is admittedly not very far away from this rationale, by providing incentives, support and subsidies to farmers according only to the environmental state of their farms and their area, and the quality and safety of the product, regardless of productivity. But, technically speaking, what is proposed here has a somewhat different underpinning and places emphasis on the systematisation of criteria based on cultural heritage.[39]

Already there seem to be some examples from national initiatives. In Sweden a programme was launched in 1995 with the primary aim of conserving biodiversity and cultural heritage values in the agricultural landscapes.[40] In England, the two largest agri-environment schemes are also aiming in the same direction. Environmentally Sensitive Areas were launched in 1987 and the Countryside Stewardship Scheme in 1991.[41] But even within the EU, the EQULFA project launched in 1996 for Less Favoured Areas of the EU seems to be exactly in line with the above recommendations.[42]

Though many gaps still exist in knowledge as well as a variety of technical issues according to the different social and institutional situations in each country, this is perhaps the safest and most effective way to provide incentives and support in practice for the conservation of the European natural heritage. By implementing this idea, conceptual uncertainties regarding natural heritage per se would be skipped, while the science of conservation biology would continue unobstructed to support the whole process.

In practical terms, however, this would require tackling the challenge of much closer collaboration and integration between scientific institutions, NGOs and administrative authorities related to farming, culture and the environment—something that has not been achieved to a satisfactory degree so far within many countries.

Two different strategies are suggested in this paper:

(1)  Systematise, elaborate, understand and describe the essence of the old farming or traditional systems used across Europe, make a typology of them and the respective spatial distribution of each type, promote them and establish them officially as the 'farming cultural heritage' of Europe, either through a convention or an EU directive or similar. The elements of this intangible cultural heritage must acquire names and identity if we want them to be acknowledged by the public and bureaucracy.

(2)  Based upon this inventory, prepare a regionally specific area or farm evaluation matrix and organise a point system through which subsidies in the agricultural sector could be distributed. Points could not only be allocated to individual farmers but additional points could be foreseen for the total effect achieved in specific areas, which could be redistributed to farmers or other responsible bodies.

In other words, use support of the cultural heritage to achieve the conservation of natural heritage.

## Notes

[1]    See the discussion in Callicott et al., 'Current Normative Concepts in Conservation'; and Lawton, 'The Science and Non-science of Conservation Biology'.

[2]    For a discussion on the term 'natural' see Götmark, 'Naturalness as an Evaluation Criterion in Nature Conservation'; and Machado, 'An Index of Naturalness'.

[3]    Lawton, 'The Science and Non-science of Conservation Biology'.

[4]    It is not strange, therefore, that in the Greek vocabulary there is no exact equivalent of the word 'conservation' and 'conservationist', and no relevant term has been coined so far.

[5]    Ahmad, 'The Scope and Definition of Heritage', 292–300.

[6]    Hobbs, in Blondel and Aronson, *Biology and Wildlife of the Mediterranean Region*.

[7]    Blondel and Aronson, *Biology and Wildlife of the Mediterranean Region*.

[8]    Myers (1988, 1990), in Blondel and Aronson, *Biology and Wildlife of the Mediterranean Region*.

[9]    Médail and Quézel, 'Biodiversity Hotspots in the Mediterranean Basin', 1510–13.

[10]   Médail and Quézel, 'Hot-spots Analysis for Conservation of Plant Biodiversity in the Mediterranean Basin,' *Annals of the Missouri Botanical Garden* 84 (1997): 112–27, in Médail and Quézel, 'Biodiversity Hotspots in the Mediterranean Basin'.

[11]   Hegemeijer and Blair (1997), in Covas and Blondel, 'Biogeography and History of the Mediterranean Bird Fauna'.

[12]   Blondel and Aronson, *Biology and Wildlife of the Mediterranean Region*.

[13]   Ibid.

[14]   Ibid.

[15]   Pienkowski 'Conservation of Biodiversity by Supporting High-nature-value Farming Systems'.

[16]   Blondel and Aronson, *Biology and Wildlife of the Mediterranean Region*, 264.

[17]   Warren, 'Conservation Biology and Agroecology in Europe'; Blondel and Aronson, *Biology and Wildlife of the Mediterranean Region*; Grove and Rackham, *The Nature of Mediterranean Europe*.

[18]   IUCN, *Parks for Life*.

[19]   Blondel and Aronson, *Biology and Wildlife of the Mediterranean Region*.

[20]   IUCN.

[21]   Di Castri, in Blondel and Aronson, *Biology and Wildlife of the Mediterranean Region*.

[22]   McCracken et al., 'The Importance of Livestock Farming for Nature Conservation'.

[23]  IUCN.
[24]  McCracken et al., 'The Importance of Livestock Farming for Nature Conservation'.
[25]  Ibid.
[26]  Lawton, 'The Science and Non-science of Conservation Biology'.
[27]  Rackham, 'Conservation Theory and Practice'.
[28]  Tella et al., 'Conflicts between Lesser Kestrel Conservation and European Agricultural Policies as Identified by Habitat Use Analyses'.
[29]  Perrings et al., 'Biodiversity in Agricultural Landscapes'.
[30]  Cahlin, 'Integrating Agriculture and Environment in Sweden'.
[31]  Baldock, 'Initial Conclusions of the 5th European Forum on Nature Conservation and Pastoralism'.
[32]  Grove and Rackham, *The Nature of Mediterranean Europe.*
[33]  However, the recent effort within the framework of the Council of Europe—titled the 'European Landscape Convention'—needs special mention, though it has not started bearing fruit yet (www.coe.int/EuropeanLandscapeConvention).
[34]  Mader, 'Wildlife in Cultivated Landscapes'; Ostermann, 'The Need for Management of Nature Conservation Sites Designated under Natura 2000'; Hopkins, 'Achieving the Aims of the EC Habitats Directive'.
[35]  McCracken et al., 'The Importance of Livestock Farming for Nature Conservation'.
[36]  Pienkowski, 'Conservation of Biodiversity by Supporting High-nature-value Farming Systems'.
[37]  This change has led either to intensification, especially in lowland, productive, soil-rich areas, or to abandonment in Less Favoured Areas (LFAs), particularly remote regions, mountain areas and islands. Both have resulted in landscape change, homogenisation and impoverishment of biodiversity. Although it is well documented that the so-called 'traditional' or old farming agro-pastoral practices favour high biodiversity and valued landscapes, there is no clear definition of what constitutes traditional in the modern world.
[38]  Bignal, 'Using an Ecological Understanding of Farmland to Reconcile Nature Conservation Requirements, EU Agriculture Policy and World Trade Agreements'.
[39]  http://ec/europa.eu/agriculture
[40]  Cahlin, 'Integrating Agriculture and Environment in Sweden'.
[41]  Ovenden et al., 'Agri-environment Schemes and their Contribution to the Conservation of Biodiversity in England'.
[42]  Fischer and Abbadessa, 'Sustainable Social and Environmental Quality in Less Favoured Areas'.

# References

Ahmad, Y. 'The Scope and Definitions of Heritage: From Tangible to Intangible'. *International Journal of Heritage Studies* 12, no. 3 (2006): 292–300.
Baldock, D. 'Initial Conclusions of the 5th European Forum on Nature Conservation and Pastoralism, 17–21 September 1996'. In *Mountain Livestock Farming and EU Policy Development: Proceedings of the 5th European Forum on Nature Conservation and Pastoralism, 18–21 September 1996*, edited by A. Poole, M. Pienkowski, D. I. McCracken, F. Petretti, C. Brédy and C. Deffeyes. Isle of Islay: European Forum on Nature Conservation and Pastoralism, 1998.
Bignal, E. M. 'Using an Ecological Understanding of Farmland to Reconcile Nature Conservation Requirements, EU Agriculture Policy and World Trade Agreements'. *Journal of Applied Ecology* 35 (1998): 949–54.
Blondel, J. and J. Aronson. *Biology and Wildlife of the Mediterranean Region.* Oxford and New York: Oxford University Press, 1999.
Cahlin, G. 'Integrating Agriculture and Environment in Sweden'. In *Mountain Livestock Farming and EU Policy Development: Proceedings of the 5th European Forum on Nature Conservation and*

*Pastoralism, 18–21 September 1996*, edited by A. Poole, M. Pienkowski, D. I. McCracken, F. Petretti, C. Brédy and C. Deffeyes. Isle of Islay: European Forum on Nature Conservation and Pastoralism, 1998.

Callicott, J. B., L. B. Crowder and K. Mumford. 'Current Normative Concepts in Conservation'. *Conservation Biology* 13, no. 1 (1999): 22–35.

Covas, R. and J. Blondel. 'Biogeography and History of the Mediterranean Bird Fauna'. *Ibis* 140 (1998): 395–407.

Fischer, G. E. and V. Abbadessa. 'Sustainable Social and Environmental Quality in Less Favoured Areas'. In *Mountain Livestock Farming and EU Policy Development: Proceedings of the 5th European Forum on Nature Conservation and Pastoralism, 18–21 September 1996*, edited by A. Poole, M. Pienkowski, D. I. McCracken, F. Petretti, C. Brédy and C. Deffeyes. Isle of Islay: European Forum on Nature Conservation and Pastoralism, 1998.

Götmark, F. 'Naturalness as an Evaluation Criterion in Nature Conservation: A Response to Anderson'. *Conservation Biology* 6, no. 3 (1992): 455–58.

Grove, A. T. and O. Rackham. *The Nature of Mediterranean Europe: An Ecological History*. New Haven and London: Yale University Press, 2001.

Hopkins, J. 'Achieving the Aims of the EC Habitats Directive: Links with the Reform of Agricultural Policies in Mountain Areas'. In *Mountain Livestock Farming and EU Policy Development: Proceedings of the 5th European Forum on Nature Conservation and Pastoralism, 18–21 September 1996*, edited by A. Poole, M. Pienkowski, D. I. McCracken, F. Petretti, C. Brédy and C. Deffeyes. Isle of Islay: European Forum on Nature Conservation and Pastoralism, 1998.

IUCN Commission on National Parks and Protected Areas. *Parks for Life: Action for Protected Areas in Europe*. Gland and Cambridge: IUCN, 1994.

Lawton, J. 'The Science and Non-science of Conservation Biology'. *Oikos* 79, no. 1 (1997): 3–5.

Machado, A. 'An Index of Naturalness'. *Journal of Nature Conservation* 12 (2004): 95–110.

Mader, H.-J. 'Wildlife in Cultivated Landscapes: Introduction'. *Biological Conservation* 54 (1990): 167–73.

McCracken, D., M. Pienkowski, E. Bignal, D. Baldock, C. Tubbs, N. Yellachich, H. Corrie and G. van Dijk. 'The Importance of Livestock Farming for Nature Conservation'. In *Mountain Livestock Farming and EU Policy Development: Proceedings of the 5th European Forum on Nature Conservation and Pastoralism, 18–21 September 1996*, edited by A. Poole, M. Pienkowski, D. I. McCracken, F. Petretti, C. Brédy and C. Deffeyes. Isle of Islay: European Forum on Nature Conservation and Pastoralism, 1998.

Médail, F. and P. Quézel. 'Biodiversity Hotspots in the Mediterranean Basin: Setting Global Conservation Priorities'. *Conservation Biology* 13, no. 6 (1999): 1510–13.

Ostermann, O. P. 'The Need for Management of Nature Conservation Sites Designated under Natura 2000'. *Journal of Applied Ecology* 35 (1998): 968–73.

Ovenden, G. N., A. R. H. Swash and D. Smallshire. 'Agri-environment Schemes and their Contribution to the Conservation of Biodiversity in England'. *Journal of Applied Ecology* 35 (1998): 955–60.

Perrings, C., L. Jackson, K. Bawa, L. Brussard, S. Brush, T. Gavin, R. Papa, U. Pascual and P. De Ruiter. 'Biodiversity in Agricultural Landscapes: Saving Natural Capital without Losing Interest'. *Conservation Biology* 20, no. 2 (2006): 263–64.

Pienkowski, M. W. 'Conservation of Biodiversity by Supporting High-nature-value Farming Systems'. *Journal of Applied Ecology* 35 (1998): 948.

Rackham, O. 'Conservation Theory and Practice' [Book review]. *Trends in Ecology and Evolution* 6 (1991): 303–4.

Tella, J. L., M. G. Forero, F. Hiraldo and J. A. Donazar. 'Conflicts between Lesser Kestrel Conservation and European Agricultural Policies as Identified by Habitat Use Analyses'. *Conservation Biology* 12, no. 3 (1998): 593–604.

Warren, J. T. 'Conservation Biology and Agroecology in Europe'. *Conservation Biology* 12, no. 3 (1998): 499–500.

# 'One and the Same *Historic* Landscape': A Physical/Cultural Perspective

John Sheail

Fragments of what are now obsolete physical environments have been lovingly preserved, and even restored as relics of times gone by. Their preservation may be costly, both directly and for their blocking new development. The often fiercely contested defence of such remnants reflects an affluence which, on the one hand, threatens them with large-scale change and, on the other, confers a desire and ability to retain them. Such remarks were made by the American writer Kevin Lynch in relation to the preservation of buildings and sets of buildings. His purpose was to highlight two things, namely the dynamics of the physical environment and, secondly, how that physical world shapes our image of the passage of time.[1]

By way of modest commentary, the present paper shifts the focus to the dynamics of the more natural world of wild plant and animal life. As the North American environmental historian Thomas Dunlap writes in his volume *Saving America's Wildlife*, its history encompasses more than natural history. It matters a lot what humans think and do. That sentiment has been so well heeded as to cause the historical ecologist Oliver Rackham to protest that historians are fixated with how people perceive and misconceive the world in which they live. A first step is surely to establish 'what they were attitudinising about'.[2] In truth, these insights are required for any comprehensive knowledge and understanding of the past, whether sought and communicated for their own intrinsic fascination or as a key to managing the present and making some prediction of the future. The different disciplines have to be recognised for what they are. Peter Fowler wrote as long ago as 1970 in the pages of *Antiquity* of welcome signs of a convergence between archaeology and ecology in developing common research and conservation interests. They were essentially specialisms within the same environmental field. In that sense, neither had been required to re-invent itself to take its place among the so-called environmental sciences.[3]

The present paper will pursue the theme of self-consciousness in illustrating how an institutional framework emerged for the study and custodianship of the 'living heritage', the enquiries consequently made by ecologists as to how such species and their respective communities functioned and, more specifically, their endeavours, through historical ecology, to bring greater rigour to the definition of 'naturalness'.

## 'Self-conscious' Ecology

The environment can be said to extend from the Van Allen Belts to deep geology; 'landscape' for the purposes of this paper refers to that part within the immediate human experience. As Claire Elizabeth Campbell has written by way of introduction to her study of the Georgian Bay area of Lake Huron in Canada, the very word 'landscape' implies broad analysis in the sense of seeing everything in context, whether thematic, spatial or functional. The widest range of perspectives must be integrated within a single story. Those stories have drawn on a century of scholarship. The landscape school of the 1920s as emerged through historical geography rejected the environmental determinism as assumed nature simply impacted upon society. By inverting that relationship, culture was found to be the agent, the natural area became the medium, and the cultural landscape the result.[4] Yet even this hardly describes the spatially intricate and temporally fluid relationships to be found at the local, regional and more global scales of such environmental histories. The human agency is exaggerated. As Ted Steinberg has written, nature was not nearly so passive, often upsetting human design to the point of redirecting historical events.[5] Such environmental historians have come to perceive people and nature as being so interrelated and dialectical that each is constantly affecting and reshaping the other. Human activity is constrained, even when pushing at the environmental limits.[6]

Such a sequence of self-questioning as to the nature of the people/nature relationship raises questions as to the self-awareness of the contributory sciences. As Robert McIntosh has written in *The Background of Ecology*, it is one thing to trace the origins of concepts and ideas which now have currency, but it is quite another to discover when and why ecologists, say, first thought of themselves as pursuing ecology. Such self-awareness in ecology seems to have emerged in the 1890s, as an interaction of natural history and physiology. It was, as McIntosh writes, not the much maligned and misrepresented 'stamp collecting' type of natural history. It was rather the natural history as promoted by Buffon and Humboldt, as held that 'observations' were not really interesting, except where they led to 'general ideas'. Humboldtian science involved accurate measurement, the development and adoption of new tools for study, and practical application out of doors.[7]

The New Zealand botanist Leonard Cockayne offers examples of the combination of influences as made for self-awareness. It might be assumed that Cockayne's fascination for plant morphology and organography simply led him into the 'unrecognized and unlabelled' field of ecology. But there was personal motivation too in his seeking 'more accurate knowledge regarding the maximum and minimum requirements of each economic plant and its behaviour when growing with other plants and animals'.[8] Ecology might not only provide employment, say in horticulture, but in 'the preservation of scenery' which, as Cockayne argues, could only be wisely effected through knowledge of what is to be protected, how it has arisen, and how it is threatened. His same intense love of the primitive vegetation of New Zealand, as had caused him to develop an unrivalled knowledge of the pre-European natural environment, had also impelled him to become an authority on how it was threatened by the introduction of exotic plant and animal species. Through self-conscious pursuit of the science of ecology, Cockayne found the means by which he might help instil a national consciousness of the distinctiveness of the New Zealand landscape and a pride in what might be accomplished by way of its promotion, both domestically and in terms of international recognition.[9]

Ecologists are bound to be key informants on the landscape. Sydney Maughan, the author of *Earth's Green Mantle*, observed in the late 1930s how 'the love of fine landscapes' comes largely through love of good 'plantscape'. Almost all the Earth's habitable surface has some kind of plant covering.[10] Arthur G. Tansley, the founding figure of British ecology, narrowed his focus to *Britain's Green Mantle: Past, Present and Future* in writing his summary volume of some 50 years of British research in 1949. Plant ecologists, joined by animal ecologists in the 1920s, had, through survey and experiment, set out to understand the impacts of soils and climate on natural vegetation. They found most vegetation to be 'semi-natural, in the sense of the plants being considerably affected by human activity, but not to the extent of their being actually planted as crops.[11] Through such a threefold distinction of natural–semi-natural–artificial, ecologists gave early recognition to human usage as an integral part of what Tansley called *Our Heritage of Wild Nature*. This was the title Tansley gave to a book published by Cambridge University Press at the end of the Second World War, with the subtitle *A Plea for Organised Nature Conservation*.[12]

That phase of post-war recovery offered numerous examples of the compelling need for objectivity in the pursuit of scientific inquiry, the motivation for such enquiry being typically personal and therefore, to a degree, socially driven, as Tansley himself freely acknowledged. The planning for post-war reconstruction had begun almost as the Second World War broke out. As early as 1940 the architect and town planner Thomas Sharp wrote of the need '*now* to consider the basis of our future plans with regard to the environment in which we live'. There was economic need for 'landscape-adaptation' (as he called it), and also need to make the landscape a more comfortable and satisfying place to live in. It was entirely understandable to cry 'Preserve the countryside', which had suffered so much damage, but the countryside was more than scenery. It was a place for social and economic activity. To arrest such dynamic change would be to destroy it. The countryside had to evolve but, as Sharp expressed it, it must change in a 'directed' manner.[13]

It was to stimulate such directive effort that Tansley wrote *Our Heritage of Wild Nature*, offering guidance on how 'organised nature conservation' might be achieved. Such prescription was considerably extended by an official post-war enquiry, the Wild Life Conservation Special Committee, with Tansley as its de facto chairman. Its report *Conservation of Nature in England and Wales*, of July 1947, pressed for the government to take a general responsibility for the flora and fauna of the country, as well as for the protection of features of geological and physiographical interest. As the Committee reasoned:

> Man is potentially the most destructive of animal species, and through ignorance and neglect frequently destroys things of great value to his own material, mental and spiritual advance.

A fuller and more intelligent use of the natural resources of the country would bring benefit to agriculture, forestry, game preservation, and fisheries, as well as to water supply, drainage, quarrying and such other aspects of civil engineering which called for physiographical and geological knowledge. It would be 'a most profitable business transaction for the State', over and above the benefit which could not be quantified in terms of money.[14]

The recommendation led to the establishment of the first body of its kind in the world, a Nature Conservancy, a research council in all but name, but also equipped to acquire and manage nature reserves, act as an expert advisory body, and to undertake such research as relevant to those responsibilities.[15] There has been much subsequent criticism of what has been perceived as 'the Great Divide' between nature conservation within the science sector of government and the scenic-amenity and outdoor-recreational aspects as remained in the land-use planning sector of government. Such criticism has exaggerated historically the common ground between them. The ecologists who largely staffed the Conservancy felt considerably greater affinity with the voluntary natural history and nature-conservation bodies. Their pressure upon government helped secure, for example, the much needed resources as enabled the Conservancy to identify the side-effects of pesticides on wildlife and game life, well before the publication of Rachel Carson's *Silent Spring* in February 1963. At a time of such lethargy among the voluntary amenity and recreational organisations, it was the nature conservation bodies that essentially instigated the 'Countryside in 1970' conferences, which prepared the ground for the British component of European Conservation Year in 1970 and the UN Stockholm conference of 1972, and that have come to be perceived as such institutional benchmarks of the so-called 'environmental revolution' of the late twentieth century.[16]

Paradoxically, the Nature Conservancy was abolished in 1973, as part of wider changes in the organisation of government research and development. Its research component remained within a considerably larger Natural Environment Research Council. The advisory and reserve functions formed a Nature Conservancy Council, a grant-aided body of executive government, the Department of the Environment. Although 'the Split' was deeply unpopular among conservationists at the time, it also acknowledged how ecologists were no longer so dependent upon nature conservation for research openings. There was similarly confidence among nature conservation bodies as to their greater understanding of 'natural' processes. And as for government, it might be argued that ecologists had successfully persuaded ministers that nature conservation was too important to be left to scientific direction![17]

## Conservation Research—Beginnings

Such institutional development owed much to gifted communicators, who contributed, for example, to the post-war 'New Naturalist' series of volumes. They sought to recapture the enquiring spirit of the old naturalists, putting across the results of modern scientific research in a way that instilled 'the natural pride of the British public in its native flora and fauna'.[18] In looking at British wildlife afresh, they recognised 'human husbandry and exploitation of natural resources as facets of nature'.[19] Such novelty of approach was conveyed especially by Richard Fitter, in his volume on London's natural history. It was a history of a great human community in terms of 'the animals and plants it had displaced, changed, moved and removed, introduced, conserved, lost or forgotten' in the course of such human activity as caring for the city's parks and gardens, the excavation of building materials, water supply and refuse disposal, food-getting, sport, and most topically the impacts of the recent bombing of the city.[20]

Such a broad sweep of natural history, which found increasing outlet through film and television, was informed by a resurgence of fieldwork and projected by the often startling discoveries made. It had been assumed that the Broads, the water bodies of the east Norfolk river valleys, were natural. Joyce Lambert, a plant ecologist, made a series of closely spaced cores through the peat in 1952 and found unmistakable signs of peat-digging. The sides of the broads were near perpendicular, and the floors flat. There was obvious puzzlement as to how they could have been excavated in the regularly inundated floodplains.[21] A further type of fieldwork ensued, by an historical geographer who found, from the archival sources in local record offices, that substantial amounts of peat were dug for fuel from at least the early 12th century. The increasing vulnerability of the area to flooding brought such exploitation to an end by the close of the 14th century.[22] Such discovery did not diminish the value of the nature reserves and Sites of Special Scientific Interest, already established in Broadland. Rather, it emphasised the extent to which past human use and management have contributed to the evolution of even those landscapes most cherished for their native flora and fauna.

The point was well taken by Eric Duffey, the Nature Conservancy's regional officer, who had assisted Lambert in her fieldwork and took charge of the research section of the Conservancy's 'applied experimental station' at Monks Wood in the early 1960s, and intended to pioneer the techniques required to manage plant and animal communities explicitly for wildlife conservation. Trials began on nature reserves using farmstock, and such cutting, mowing and burning treatments as might replicate how the communities had been formed and previously managed.[23] An advertisement was placed in the summer of 1967 for an historical geographer to join 'biologists working on ecological problems of nature reserve management'. Because many reserves occupied land that had been used over long periods of time, the 'further job particulars' explained how 'biological research on present-day plant and animal communities is greatly helped if a knowledge of the ecological history is available'.

As Leonard Cockayne might have perceived earlier, the Conservancy's priority was to discover how representative its nature reserves were of the plant and animal communities that had once existed. Cockayne's contemporary, George M. Thomson, confesses in his volume *The Naturalisation of Animals and Plants in New Zealand* to the difficulty of making such historical enquiry in a country as recently settled as New Zealand.[24] The challenge was all the harder in a country so long and intensively settled as Britain, and yet it had to be met if even the most tentative response was to be given to such questions as the historical impact of such alien species as the rabbit. Given the animal's significance on the Conservancy's reserves, immediately prior to the outbreak of the virus myxomatosis in the mid-1950s, an obvious need was to establish when and where, and how and why, the rabbit became so ubiquitous and abundant. What were the ecological consequences of its spread to the different habitats? Some insight was obtained from improved biological understanding of how the species adapts to new environments. Investigation was made of the archaeological remains of the grounds and in buildings of its earlier protection for fur and meat.[25] And searches might be made of published and archival sources.[26] To cite an instance of the latter, the *Memories* of Lady Gordon Cumming describe how rabbits had never been seen in the woods of her family's estate on the Findhorn River of Scotland, until an English gamekeeper was engaged in 1816, who set about destroying the natural predator population.[27]

Research has optimally proceeded at a range of scales. The Cambridgeshire and Isle of Ely Naturalists' Trust acquired Hayley Wood in 1962, as the largest surviving oak-ash wood in west Cambridgeshire and, more particularly, for its very large population of the oxlip (*Primula elatior*). Its richly documented history suggested that the 49 ha site had been wooded since at least 1251. The Trust's publication of a pioneering account of the reserve's social history and present-day ecology stimulated more general investigation, the principal author, Oliver Rackham (a plant physiologist and ecologist) going on to publish *Ancient Woodland: Its History, Vegetation and Uses in England*, which drew upon the archives and field evidence of many further woodlands, principally in eastern England. There followed Rackham's *The History of the Countryside*.[28]

Truly interdisciplinary research can never be taken for granted. Situations can arise analogous to what a behavioural ecologist once witnessed on a particularly rich salmon run in Alaska. The bears and fishermen almost rubbed elbows as they fished for the same prey by different methods. They were so entirely focused on their fishing as hardly to notice the other.[29] Alongside such fixation of the respective disciplines, Hugh Trevor Roper found most professional historians specialised in the knowledge that, so armed, 'they can comfortably shoot down any amateurs'. As for themselves, they knew the best form of defence was to 'keep prudently within their own frontiers'.[30] In truth, the Historical Ecology Discussion Group, which met at the Nature Conservancy's Monks Wood Experimental Station from the late 1960s, received much encouraging support.[31] As William Hoskins expressed it in his volume of essays that accompanied the television series *Landscapes of England*, the same question was asked of every landscape: 'How did it come to be like this?'[32] The ensuing collaborative investigations were, for the British part, quite unpretentious.

The convergence of research interests in landscape development proved no passing phase of academia. There has been an enduring interest in hedgerows ever since Max Hooper at Monks Wood suggested the diversity of their shrub species might be related to their age and manner of creation.[33] Conservation measures and, more recently, statutory controls gave added impetus to assessing their wildlife value. Of the many publications, those utilising both field and archival evidence have most vigorously asserted the relationship between species composition and age. The authors of a case study of the South Gower found it was a reliable guide to succession, noting how Hooper had never claimed it to be either a method for close dating or one that was universally applicable.[34]

For three decades environmental historians (as they came to be called) have explored the people/nature relationship, endeavouring to obtain what Timothy Silver called the 'earth's-eye view of the past', giving as much attention to wild plant and animal life (from the redwood to the micro-organism), as to the more immediate preoccupations of humankind.[35] Such histories are, however, severely constrained by what previous generations chose to record by way of literary, archival and oral memory. The geographers Janet Hooke and Roger Kain[36] were among the first to identify in the form of a published guide what relevant sources were available for the British physical environment, emphasising how much of the evidence remained preserved within the landscape itself. The skills of a civil engineer considerably advanced understanding of the significance of the cast iron column, the Holme Fen post, standing within the Holme Fen National Nature Reserve of Cambridgeshire. Landowners had expected there to be large-scale surface subsidence following the drainage of the peat soils of that part of the fenland in 1850. They drove the post vertically down to the underlying clay as a measure of that subsidence. Its present-day exposure records four phases of surface lowering, as correspond with the initial drainage, installation of more powerful steam pumps in 1877, diesel pumps in 1924, and electric power in 1961.[37]

A striking example of the value of field evidence, in confirming what might be learned from the printed literature, was an examination of sand deposits in the East Anglian Breckland. There seemed every reason to be sceptical of a paper published in the *Philosophical Transactions of the Royal Society of London* in 1668, describing a 'sand-floud', as reputedly destroying cornfields and houses in Santon Downham, a village 8 km distant.[38] Opportunity was taken, through a programme of optically stimulated luminescence and radiocarbon dating, to examine the extant inland dunes of

Wangford Warren, together with such deposits found in Santon Downham. Of the five sand-depositional phases identified since 6,500 BP, the most recent dated around 400–335 BP, and another 200–30 years BP. Whilst the authors emphasise the local complexity of such aeolian activity, the field evidence appears to confirm a 17th-century 'sand-flood' that could be attributed to low temperatures and high rabbit and sheep populations reducing the vegetation cover. The final phase of activity (between 200 and 30 years BP) might reflect the expansion of arable farming, combined with increased temperatures. Both episodes corresponded with much inland and coastal dune building in other parts of north-west Europe.[39]

## Research Conservation—its Wider Application

A 'siege' mentality gripped British nature conservationists in the 1970s. Field and archival surveys were commissioned to highlight the enormity of what was being irrevocably destroyed through agricultural improvement and afforestation. The Countryside Commission initiated a survey of vegetation change in 12 upland landscapes, totalling 74,000 ha, in England and Wales. Surveyors found the mosaic of moorland types to be much more than a function of altitude and rock type. Over 10% of the moorland—usually upon the moorland fringe—had been converted to farmland or forest over the previous 200 years, and a further 2,500 ha of farmland had reverted over that period to moorland. Whilst reclamation brought rapid change, reversion, say through rough pasture to scrubby heath, was so slow that, even after 100 years, it might still be noticeably different from those parts never improved. Such explanation might be found in how grazing animals were drawn to the residual effects of previous soil improvement, their preferential cropping, and therefore their urine and dung, sustaining the more rapid turnover of nutrients.[40]

It was this kind of field observation that encouraged ecologists of the erstwhile research branch of the Nature Conservancy (now the Institute of Terrestrial Ecology) to press for experimentation to develop a more exact understanding of the impacts of modern husbandry practices, as a key to mitigating their most damaging effects. They pressed for trials whereby the drainage pumps were turned off, the drains closed, and the wildlife of wetter conditions encouraged. Derek Ratcliffe, Chief Scientist of the Nature Conservancy Council, argued that, even if such restoration of wetlands was practical, it would be extremely expensive.[41] Only when the still-important wetland sites and species had been protected, or actually lost, should monies be invested in attempts to re-create biotopes *de novo*. The ecologists' proposals became, however, part of a considerably larger agenda, as the changing circumstances of farming made proposals for such management and restorative work increasingly relevant to the wider countryside. The Ministry of Agriculture recognised that farming (like any business) must adjust to changing market demand for its products, as well as to mounting anxieties about the effects of its activities on the environment.[42] There was public resentment at the cost of subsidising production of foodstuffs, many of which were now in surplus under the European Community's Common Agricultural Policy.

Agriculture ministers were required, under the Agriculture Act of 1986, to strike a reasonable balance between 'the promotion and maintenance of a stable and efficient agricultural industry' and the other economic, social, conservation and recreational interests of rural areas. Whilst only a very small proportion of the annual agricultural-support budget was transferred to countryside management schemes, there was at least the possibility of 'pump-priming' monies being used to promote what the Royal Society for the Protection of Birds described as a more integrated approach to farming and the environment.[43] Wildlife protection became not so much a concession as a component part of farming operations. Sixty-five areas of the UK were designated as Environmentally Sensitive Areas by 1995, where grants were available to those farmers going beyond 'normal responsible husbandry practices, so as to manage their land in ways that protected natural beauty, wildlife, or buildings and other features of archaeological, architectural or historic interest'.

The research initiative for enhancing the wildlife of the wider countryside had passed to the Agriculture Departments. However, ministers were in a position analogous to the Nature Conservancy in the 1950s. The policy instruments were in place, but it was far from clear whether there was the expertise to ensure the desired goals could be met. Ceasing damaging agricultural practices would not, of itself, restore the target wildlife communities. 'Natural' reversion was far too slow and unreliable.[44] Prescriptions were required for habitat restoration that were administratively simple yet tailored to meet local circumstances. Conservation bodies feared that such notions of 'creative conservation' would weaken their case for protecting sites. Developers would argue that anything destroyed could be re-created elsewhere. The ecologists actually engaged in such restorative work profoundly disagreed. They emphasised the scientific and technical difficulties of establishing closely prescribed, species-rich plant communities, let alone the re-introduction of their associated invertebrate and vertebrate fauna. The intention was to enhance what might otherwise disappear, but there was neither the competence nor the wish to create facsimiles of some previously 'pristine' state—to 'fake nature', as one social scientist put it.[45]

## The Shifting Ecological Context

Oliver Rackham concluded his 'New Naturalist Library' volume *Woodlands*, the 100th in the series, by citing the experience of the recent past to illustrate how hazardous it was to predict the future. Who would have thought, in the seemingly precarious years of the late 1960s and 1970s, that many of the threats to woodland trees and their understorey would have diminished, if not disappeared altogether, a quarter-century later? Losses there have been, some grievous, but, as Rackham contended, the priority has become one of pausing, thinking, and ensuring that the detail for future conservation measures is right.[46] Such respite (if it proves to be) is all the more reason for ensuring that there is a more informed response in deciding 'What pieces of the environment warrant some form of intervention and protection.' As Kevin Lynch has written in relation to the conservation of buildings, choices have to be made. Everything in the landscape becomes 'historical' and one cannot keep everything.[47] The realisation that it is 'One and the same *Historic* Landscape', where the physical and cultural dimensions are often difficult to distinguish, means all are implicated in decision making.

The challenge in exercising such choice arises both from the remarkable diversity of the landscape and from the range of perceptions brought to bear. The Nature Conservancy offered one approach in its choice of the 100 or so national nature reserves designated in the first 10 years. Lullington Heath on the South Downs of southern England was typically chosen in 1956 for its scientific importance. It was one of the largest tracts of unploughed chalk heath in Britain. Ecologists were intrigued at how plants requiring acid soil (such as ling and bell heather) could be intermingled with others needing a limey soil. Cambridge plant ecologist Peter Grubb made full use of the reserve in investigating what proved to be inter-species competition in an essentially open grass-heathland, grazed by sheep, cattle and rabbits.[48] The reserve manager, on the 50th anniversary, emphasised a quite different management objective, namely to accommodate as wide a variety of habitat as possible. He rejoiced at how his use of a goat-herd, a few ponies and rabbits had created a landscape which looked to him like an African savanna. Further perceptions may be confidently expected following the absorption of these reserves in 2006 into Natural England, a body encompassing nature conservation, scenic amenity and outdoor recreation. The name was reputedly chosen as arousing the least objection of all those canvassed.

The adoption of the word 'natural' raises the question even more starkly as to the range of perceptions brought to bear on the landscape. George Peterken remarked, a decade earlier, on how conservationists (in Britain at least) perceived 'naturalness' as conveying everything ideal by way of diversity, grandeur, health and vigour. For Peterken, such an assumption was severely challenged by a visit to the Yellowstone National Park. The younger woodland growth, where it had not been affected by fire or disease, looked as dense and monotonous as any plantation so despised by conservationists in upland Britain. And to the British eye, the older growth of Yellowstone was untidy, shabby and unhealthy, despite such ravages as fire and disease being part of that ecosystem's natural order. The message for Peterken was to recognise that 'naturalness' was no absolute measure, but rather a continuous variable.[49] In a very real sense, all woodland was semi-natural in that none was entirely free of human influence, but that even the most vigorously managed plantation had some wildlife.

Such reassessment of 'naturalness' removes a large measure of certainty—so beloved of the more belligerent conservationists.[50] The definition of 'naturalness', as the *degree* of disturbance, can, however, advance dialogue considerably with other interests. The British Ministry of Defence was once a principal bogeyman of conservationists for its occupancy and treatment of such areas as Salisbury Plain Training Area in south-central England. The 38,000 ha are the most continuous expanse of agriculturally unimproved chalk grassland in north-west Europe, with plant species populations of up to $30–40/m^2$. The military occupancy from 1897 onwards preserved the Plain from the most recent, intensive agricultural practices. The grasslands have not, however, been 'fossilised', in the sense that vehicle movement and missile impacts have disturbed the vegetation cover as to create such habitats as the ephemeral pools of Imber Valley that contain internationally important populations of the Fairy shrimp. The deeply-rutted tracks mimic the conditions of the historic coach-roads across the Plain.[51] To that extent the Ministry of Defence meets its obligation of taking account of the conservation value of the Plain. The question arises, however, of when such usage becomes damaging. Where and when, on the spectrum of such levels of disturbance, might this activity impair 'naturalness' to the extent of its becoming irreversible? The exceptionally fine series of aerial photographs available for the period 1945–1995, together with satellite imagery, show that disturbance increased significantly in the 1990s. More positively, such historical information now feeds into the integrated management plans required by the Plain's managers. The trajectories enable ecologists to be more confident in prescribing the available options.[52]

Leonard Cockayne's research in New Zealand illustrated how ecology is advanced through close observation of particular circumstances. Ian Lunt and Peter Spooner, a century later, have researched the agriculturally fragmented landscapes of south-east Australia in explaining spatial variations of species populations. Rather than dismissing the remnants of habitats encompassed by the nature reserve and hedgerow-bank as simply 'noise' or 'the mess of history', the incorporation of the historical ecological evidence may considerably advance the ecologist's understanding of island biogeography theory. There is practical spin-off too in calling into question the commonplace insistence that there should be no human disturbance within nature reserves, if they are to be restored to some former pristine condition. Given the evidence for significant aboriginal and early-settler usage, such exclusion will not so much restore pristine conditions as establish entirely novel ecosystems. Historical ecology studies should be an essential part of any conservation management programme.[53]

In disseminating their research findings, ecologists are required by their respective institutions to communicate as part of the larger scientific community. Considerable emphasis is rightly placed on the peer-review process in ensuring the soundness of what is published. That process should be regarded, however, as only the beginning in relaying such research findings to those best placed to give any practical effect to what has been discovered. To secure that end, authors might be expected to choose between a numerous and diverse range of journals, publishers perhaps competing for the most scholarly and relevant. Instead, authors are increasingly required to compete among themselves in submitting papers to journals perceived by a largely faceless bureaucracy as the highest in some 'pecking order'. The most highly esteemed journals become so bloated with offerings as to define their respective core interests ever more narrowly. By its very diversity, landscape research is particularly affected. Such inhibition has become all the more serious, given the significance of its message.

## Notes

[1]   Lynch, *What Time is this Place?*
[2]   Dunlap, *Saving America's Wildlife*; Rackham, 'Prospects for Landscape History and Historical Ecology'.
[3]   Fowler, 'Old Grassland'.
[4]   Campbell, *Shaped by the West Wind*, 5, 11–12.

[5]   Steinberg, *Down to Earth*, 284–85.
[6]   Campbell, *Shaped by the West Wind*, 17–18.
[7]   McIntosh, *The Background of Ecology*, 21–27.
[8]   Hill, 'Leonard Cockayne', 444.
[9]   Cockayne, *New Zealand Plants and their Story*; Hill, 'Leonard Cockayne', 453–54.
[10]  Maughan, *Earth's Green Mantle*, 19.
[11]  Tansley, *Britain's Green Mantle*.
[12]  Tansley, *Our Heritage of Wild Nature*.
[13]  Sharp, *Town Planning*, vii, 26, 35.
[14]  Minister of Town and Country Planning, *Conservation of Nature in England and Wales*, 64–65.
[15]  Sheail, *Seventy-five Years in Ecology*.
[16]  Sheail, *An Environmental History of Twentieth-century Britain*.
[17]  Sheail, *Seventy-five Years in Ecology*.
[18]  Marren, *The New Naturalists*.
[19]  Stamp, *Nature Conservation in Britain*, x–xii.
[20]  Fitter, *London's Natural History*.
[21]  George and Jermy, 'Joyce Lambert'.
[22]  Lambert et al., *The Making of the Broads*.
[23]  Duffey et al., *Grassland Ecology and Wildlife Management*.
[24]  Thomson, *The Naturalisation of Animals and Plants in New Zealand*.
[25]  Williamson, *The Archaeology of Rabbit Warrens*.
[26]  Sheail, *Rabbits and their History*.
[27]  Gordon Cumming, *Memories*, 388–91.
[28]  Rackham, *Hayley Wood*; *Ancient Woodland*; *The History of the Countryside*.
[29]  J. Kraus, 'Menu of Grizzly Delights', *The Times Higher*, 25 August 2006, 18.
[30]  Trevor Roper, *Historical Essays*, v–vi.
[31]  Sheail, 'Hoskins and Historical Ecology'.
[32]  Hoskins, *One Man's England*.
[33]  Pollard et al., *Hedges*.
[34]  Edwards et al., 'Hedgerows and the Historic Landscape'.
[35]  Silver, *Mount Mitchell and the Black Mountains*, xiv–xv.
[36]  Hooke and Kain, *Historical Change in the Physical Environment*.
[37]  Hutchinson, 'The Record of Peat Wastage in the East Anglian Fenlands at Holme Post'.
[38]  Wright, 'A Curious and Exact Relation of a Sandfloud'.
[39]  Bateman and Godby, 'Late-Holocene Inland Dune Activity in the UK'.
[40]  Ball et al., *Vegetation Change in Upland Landscapes*.
[41]  Ratcliffe, 'Concluding Remarks', 57–60.
[42]  Ministry of Agriculture, *Our Farming Future*.
[43]  Royal Society for the Protection of Birds, *Agriculture and the Environment*.
[44]  Mountford et al., 'Reversion of Vegetation Following the Cessation of Fertiliser Application'.
[45]  Sheail et al., 'The UK Transition from Nature Preservation to "Creative Conservation"'; Elliot, *Faking Nature*.
[46]  Rackham, *Woodlands*, 558–59.
[47]  Lynch, *What Time is this Place?*, 36.
[48]  Grubb et al., 'The Ecology of Chalk Heath'.
[49]  Peterken, *Natural Woodland*.
[50]  Adams, *Nature's Place*.
[51]  Walker and Pywell, 'Grassland Communities on Salisbury Plain Training Area'.
[52]  Hirst et al., 'Assessing Habitat Disturbance Using an Historical Perspective'.
[53]  Lunt and Spooner, 'Using Historical Ecology to Understand Patterns of Biodiversity in Fragmented Agricultural Landscapes'.

# References

Adams, W. M. *Nature's Place.* London: Allen & Unwin, 1986.

Ball, D. F., J. Dale et al. *Vegetation Change in Upland Landscapes.* Cambridge: Institute of Terrestrial Ecology, 1982.

Bateman, M. D. and S. P. Godby. 'Late-Holocene Inland Dune Activity in the UK.' *The Holocene* 14 (2004): 579–88.

Campbell, C. E. *Shaped by the West Wind.* Vancouver: University of British Columbia Press, 2005.

Cockayne, L. *New Zealand Plants and their Story.* Wellington: John Mackay, 1910.

Duffey, E., M. G. Morris et al. *Grassland Ecology and Wildlife Management.* London: Chapman & Hall, 1974.

Dunlap, T. R. *Saving America's Wildlife.* Princeton: Princeton University Press, 1988.

Edwards, K., D. Leighton and P. Llewellyn. 'Hedgerows and the Historic Landscape.' *British Wildlife* 17 (2006): 260–69.

Elliot, R. *Faking Nature.* London: Routledge, 1997.

Fitter, R. S. R. *London's Natural History.* London: Collins, 1945.

Fowler, P. J. 'Old Grassland'. *Antiquity* 44 (1970): 57–59.

George, M. and C. Jermy. 'Joyce Lambert'. *Watsonia* 26 (2006): 208–9.

Gordon Cumming, C. F. *Memories.* Edinburgh: Blackwood, 1904.

Grubb, P. J., H. E. Green and C. J. Merrifield. 'The Ecology of Chalk Heath.' *Journal of Ecology* 57 (1969): 175–212.

Hill, A. W. 'Leonard Cockayne'. *Obituary Notices of Fellows of the Royal Society* 1 (1934): 443–57.

Hirst, R. A., R. F. Pywell and P. D. Putwain. 'Assessing Habitat Disturbance Using an Historical Perspective'. *Journal of Environmental Management* 60 (2000): 181–93.

Hooke, J. M. and R. J. P. Kain. *Historical Change in the Physical Environment.* London: Butterworth, 1982.

Hoskins, W. G. *One Man's England.* London: BBC (1976)..

Hutchinson, J. N. 'The Record of Peat Wastage in the East Anglian Fenlands at Holme Post'. *Journal of Ecology* 68 (1980): 229–49.

Lambert, J. M., J. N. Jennings et al. *The Making of the Broads.* RGS Research Series 3. London: Murray, 1961.

Lunt, I. D. and P. G. Spooner. 'Using Historical Ecology to Understand Patterns of Biodiversity in Fragmented Agricultural Landscapes'. *Journal of Biogeography* 32 (2005): 1859–73.

Lynch, K. *What Time is this Place?* Cambridge, MA: MIT Press, 1972.

Marren, P. *The New Naturalists.* London: HarperCollins, 1995.

Maughan, S. *Earth's Green Mantle.* London: English University Press, 1939.

McIntosh, R. P. *The Background of Ecology.* Cambridge: Cambridge University Press, 1985.

Minister of Town and Country Planning. *Conservation of Nature in England and Wales.* Cmd 7122. London: HMSO, 1947.

Ministry of Agriculture. *Our Farming Future.* London: Ministry of Agriculture, 1991.

Mountford, J. M., K. H. Lakhani and R. J. Holland. 'Reversion of Vegetation Following the Cessation of Fertiliser Application'. *Journal of Vegetation Science* 7 (1996): 219–29.

Peterken, G. F. *Natural Woodland.* Cambridge: Cambridge University Press, 1996.

Pollard, E., M. D. Hooper and N. W. Moore. *Hedges.* London: Collins, 1974.

Rackham, O. *Hayley Wood: Its History and Ecology.* Cambridge: Cambridgeshire and Isle of Ely Naturalists' Trust, 1975.

———. *Ancient Woodland: Its History, Vegetation and Uses in England.* London: Edward Arnold, 1980.

———. *The History of the Countryside.* London: Dent, 1986.

———. 'Prospects for Landscape History and Historical Ecology'. *Landscapes* 2 (2000): 3–15.

———. *Woodlands.* London: Collins, 2006.

Ratcliffe, D. A. 'Concluding Remarks'. In *Habitat Restoration and Reconstruction,* edited by E. Duffey. Huntingdon: Recreation Ecology Research Group, 1981.

Royal Society for the Protection of Birds. *Agriculture and the Environment*. Sandy: RSPB, 1991.

Sharp, T. *Town Planning*. Harmondsworth: Penguin Pelican, 1940.

Sheail, J. *Rabbits and their History*. Newton Abbot: David & Charles, 1971.

———. *Seventy-five Years in Ecology*. Oxford: Blackwell Scientific, 1987.

———. *Nature Conservation in Britain: The Formative Years*. London: The Stationery Office. 1998.

———. *An Environmental History of Twentieth-century Britain*. Basingstoke: Palgrave, 2002.

———. 'Hoskins and Historical Ecology.' In *Post-medieval Landscapes in Britain: Landscape History after Hoskins*, 3, edited by M. Palmer and P. S. Barnwell. Macclesfield: Windgather (forthcoming).

Sheail, J., J. R. Treweek and J. M. Mountford. 'The UK Transition from Nature Preservation to "Creative Conservation"'. *Environmental Conservation* 24 (1997): 224–35.

Silver, T. *Mount Mitchell and the Black Mountains*. Chapel Hill: University of North Carolina Press, 2003.

Stamp, D. *Nature Conservation in Britain*. London: Collins, 1969.

Steinberg, T. *Down to Earth*. New York: Oxford University Press, 2002.

Tansley, A. G. *Our Heritage of Wild Nature: A Plea for Organised Nature Conservation*. Cambridge: Cambridge University Press, 1945.

———. *Britain's Green Mantle: Past, Present and Future*. London: Allen & Unwin, 1949.

Thomson, G. M. *The Naturalisation of Animals and Plants in New Zealand*. Cambridge: Cambridge University Press, 1922.

Trevor Roper, H. R. *Historical Essays*. London: Macmillan, 1957.

Walker, K. and R. Pywell. 'Grassland Communities on Salisbury Plain Training Area'. *Journal of the Wiltshire Botanical Society* 3 (2000): 15–27.

Williamson, T. *The Archaeology of Rabbit Warrens*. Princes Risborough: Shire Publications, 2006.

Wright, T. 'A Curious and Exact Relation of a Sandfloud'. *Philosophical Transactions of the Royal Society of London* 3 (1668): 722–25.

# Human Heritage and Natural Heritage in the Everglades

James A. Kushlan & Eileen M. Smith-Cavros

Cultural and natural heritages of the Everglades have always been inextricably inter-twined. History demonstrates that humans have confronted nature in attempting to develop the Everglades and its surrounding areas and these attempts continue today at an ever-increasing intensity. This has resulted in a contemporary landscape that differs significantly from what would have occurred without man's presence. South Florida's human community today is defined by its intriguingly chaotic cultural diversity. In this paper, we examine the mutual evolution of local human culture and the Everglades environment, suggesting that the famously intrusive current development trends are continuous with those of previous cultures. We consider the persistent difficulty of establishing a mutually compatible existence of human and natural environment in southern Florida and suggest that the conflict is fuelled by local desires for economic benefit influenced by a sense of impermanency and access to a growing technological ability that appears to favour self-interest and short-term gains in contrast to the wider-world ethic favouring conservation of the environment.

## The Natural Heritage of South Florida

The natural heritage of south Florida is of world significance. The Everglades is one of the great wetlands of the world, originally encompassing nearly all the south Florida mainland, located at the south-eastern tip of North America. Historically, the Everglades was a marsh complex of herbaceous vegetation (dominated by razor-sharp sawgrass (*Cladium jamaicensus*)) broken intermittently by tree-covered 'islands', an environment that, around AD 1900, covered over 10,000 km$^2$.[1] This Everglades, proper, occupied the slightest of valleys, 100 km long, south of Lake Okeechobee, positioned in the centre of the Florida peninsula, reaching to the ocean along three coasts. Unlike most large tropical wetlands, it lacked rivers except at its fringes. In the rainy season, water flowed southward from the lake over a gradient of 3 cm/km to the sea. Surface water connected directly to subsurface aquifers in such quantity that water emerged under pressure at various points. The remainder of southern Florida was also part of the greater Everglades wetland environment, connected ecologically by seasonal inundation. To the north were deepwater swamps fed by Lake Okeechobee overflow. To the west, a great swamp was dominated by cypress trees (*Taxodium distichum*). To the south, extensive mangrove swamps occurred where freshwater and saltwater mixed within narrow estuaries. To the east and west, a mix of swamps and marshes melded into a narrow rock and sand ridge that lined the Florida coasts. The highest lands were the narrow ridges and islands along the east coast and the islands along the west coast. The extent of dry land varied seasonally with the subtropical rainfall pattern. On average the area received 4 m of rain annually, mostly during the rainy season of May to November. Flooded areas expanded in the rainy season. In the dry season, water depths fell, until by May much of the landscape lacked standing surface water. This environment, together covering 20,000 km$^2$, can best be thought of as a sea of shallow fresh, brackish, and saltwater marshes, swamps and bays dotted by intermittent islands, most prominently along the east and west sea coasts. Currently, the remnant core of the Everglades and its adjacent cypress swamps and mangrove swamps have been set aside from development to be used as reservoirs, flood control and water delivery systems, natural parks and reserves. Everglades National Park is perhaps the best known of the dozens of entities that make up this collection of reserved land. Reserves, though, constitute less than half of the wetland present at the turn of the 20th century.

## Human Culture in South Florida through History

Paleo-Indians were first recorded in Florida about 17,000–10,000 years BP, at a time when sea levels were as much as 30 m lower than at present.[2] The environment of the then much larger Florida peninsula was cool, dry, and windy, an effect of the glaciation of much of the continent, which was only marginally suitable for human occupation. Humans in Florida during this Paleo-Indian period were part of the North American Clovis culture, hunter-gatherers specialising in mega-fauna hunting and gathering who practised a nomadic way of life in a savannah-like environment. They occupied high ground, where the game occurred. In this dry environment, there was no Everglades. Therefore, humans occupied the land that later became south Florida before the Everglades developed.

The Archaic Indian cultures followed the Paleo-Indians. Climate changes, starting about 9,000 BP, brought about significant change to the south Florida environment, reducing the Pleistocene mega-fauna, increasing rainfall, temperatures, diversity of plant and animal life, and sea level.[3] During this Early Archaic period, human populations increased in the more benign environment and developed more sophisticated subsistence hunting, trapping, fishing, and gathering, and became less nomadic.[4] The wetter conditions that caused the transition from the Paleo-Indian hunting culture to the more versatile Archaic culture were interrupted by an extensive dry period lasting from 8,000 to 5,000 BP. Indian populations declined as conditions became harsher. People lived in small family groups near water.

Conditions changed again starting at 5,000 BP, leading to a new cultural tradition, called the Late Archaic period (5,000–2,500 BP). This period saw a wetter environment, a rise in sea level to near current stage, and the development of extensive wetland ecosystems, including the great inland wetlands of Lake Okeechobee and the Everglades. These wetlands isolated portions of south Florida from the remainder of the peninsula and the peoples there developed distinctive local adaptations.[5] By the late Archaic period, the peoples of south Florida had settled in small villages, and hunted and gathered a diverse diet of marine and freshwater animals, especially fish and turtles. In addition, they collected plant parts such as acorns for flour. In south Florida, village sites were located exclusively on islands along the coast. Hunters and gatherers used the inhospitable inland swamps only seasonally. However, like Paleo-Indians before them, they continued to be non-agricultural. A critical feature of their villages was the construction of mounds, particularly using shells to elevate home sites and burial sites above water.

The Glades period from 2,500 BP to AD 1,513 was a time of evolving local cultural adaptation to the emerging south Florida environment.[6] The Calusa, as did their predecessors, lived in large villages on high ground, especially on islands along the coast, with the largest being on Pine Island on the Gulf of Mexico coast north-west of the Glade region. Elevation was one of the crucial factors in defining a village site.

Useppa Island, enjoying the highest elevation along the Gulf coast, was occupied continuously, and with little change in customs, for 5,000 years; Mound Key, the seat of the chief of the Calusa, was a largely artificial island consisting of a 10 m high platform built of shells; and on Pine Island canals were dug for access, one of which ran 5 km.[7] Some islands were made many metres high, of shells, with internal networks of ridges, mounds, lagoons as well as seawalls and jetties, also made of shells that protected the village harbour. On Marco Island, canals ran to the centre of the island adjacent to which houses were built. Indians in this period lived not only on islands and high ground along the Gulf of Mexico but also along the southern coast, in the Florida Keys, along the rock and sand ridge east of the Everglades, and on the high ground near Lake Okeechobee.

The Calusa was a sophisticated society characterised by artistically crafted artefacts, organised hierarchies, and a region-wide political structure.[8] The major chief from Mound Key overlorded as many as 50 tributary villages along both south Florida coasts, including the Tequesta and Jobe Indians who resided along the eastern edge of the Everglades (areas later to be Miami to Palm Beaches).[9] The economy continued to be based on seasonal gathering, fishing, and some hunting. There is little evidence of agriculture along the coast. Fruits and plant materials were gathered seasonally, supplemented by trade with inland villages along Lake Okeechobee, where some root crop agriculture probably occurred. Most of the diet was marine and estuarine derived—resources that were highly seasonal. Fertility rituals developed around this seasonality—for example, the seasonal migrations of the mullet were accompanied by human sacrifice.[10] The rich food sources of the Gulf estuaries supported population growth. In 1566, Pedro Menendez de Aviles was hosted by more than 4,000 Calusa on Mound Key, suggesting the population numbered about 10,000.[11] It is estimated that by AD 800 Indians occupied all suitable living sites in south Florida that included all the available high ground. They did not live in the Everglades, which was used only for seasonal forays without permanent habitation. Dependence on marine and estuarine food sources to support the large population was also the imperative for Calusa dominance over the remaining tribes, as food redistribution was critical to population maintenance.[12]

Thus at the time of Spanish contact, south Florida was fully and sustainably populated. People used the highest ground and were able to elevate sites further by using seashells to create mounds and dikes, and to dig canals for transport trails. There was no local agricultural tradition, given the tiny amount of dry ground. The economy relied on subsistence exploitation—fishing, hunting and gathering—as ancestral economies had done for several thousand years. Much of this culture was driven by opportunities to exploit natural resources and the landscape, but at the same time, exploitation was restricted by the limited technology available. With access only to shells and hand labour it was beyond the ability of local villages further to drain, canalise and fill land. Exploitation, and the redistribution of estuarine resources, drove both the economy and local politics.

The next cultural period, which may be termed the Contact period, following first contact by Europeans, was a time of cultural extinction in south Florida. In 1513 on the east coast, and 1565 on the west coast, the Spanish began their contacts. Overall, the marshes and swamps of south Florida proved of little interest to the Spanish, who recognised the land correctly as a 'very poor land subject to inundation',[13] other than their concerns over issues such as shipwrecks, pirates, and incursions by other Europeans. In south Florida, they left little direct physical impact. However, Spanish contact did begin the annihilation of the indigenous Indians through disease and removal, first by the Spanish and later by northern Indians and Americans. Thus began a period in which south Florida was essentially depopulated, and ancient cultural traditions lost. By the mid-18th century the few remaining Indians led a poor, seasonally nomadic fishing existence. In 1763, nearly all the remaining Indians chose to leave with the Spanish when Florida was deeded to England.[14]

Then followed what may be called the Unoccupied period. South Florida was little populated for 150 years, from the first departure of the Spanish until the end of the 19th century. The environment that in any case had been little changed by thousands of years of occupancy by indigenous Indians, other than elevation of their home sites, reverted completely to its natural situation. The ancient coastal village sites were reclaimed by vegetation and the interior wetlands, which were never altered by early Indians, continued to function in their natural way. For this long period, natural processes continued unabated and the land was undisturbed by people. There are tales that when people returned they found animals that had lost their wariness of humans. It is likely that marine and estuarine stocks recovered and plant communities, mammals, and birds achieved a natural balance with the seasonal environment.

The period of re-occupancy of south Florida may be called the American period. The human-determined future of south Florida was driven by the American imperative for national expansion. Four aspects of this expansion together had profound effects locally. The first was the policy of removal of Indians from the eastern USA. The second was the opening of land for agriculture. The third was the availability of pioneers ready to seek new opportunities in newly opened lands. The fourth was the improved technology that came with the onset of the industrial revolution in America which allowed for expanded Everglades development that had been technologically impossible for early inhabitants of Florida. All of these forces came together in south Florida around 1880.

The removal policy in south-eastern North America resulted in two cultural traditions of Creek Indians migrating into Florida, who, with escaped slaves, evolved quickly into the Seminole nation.[15] After the Second and Third Seminole Wars (1835–1842, 1854–1858) remnant Seminoles retreated from their temperate forests to the Everglades. These groups thus may be considered intra-national 'immigrants', first in a rising stream of modern-day intra-national and international migrants to south Florida. One wonders whether, like many later immigrants, some Seminoles may have seen south Florida as a temporary respite. The cultural shift required of them in migrating to the Everglades was profound indeed, as the Creeks were hunters and agriculturalists with corn as their staple. But the Everglades, marked by impassable sawgrass sloughs, limited game, few fish, and seasonally fluctuating water depths, provided as little sustenance as it did for previous Indians, who had used it only seasonally. So the Seminoles in the Everglades developed a family-based culture (since the islands were small), relying on only small-game hunting, fishing, gathering and small-scale agriculture as well as on extensive trade with Americans of hunted goods for manufactured goods. For the first time since its formation, the Everglades was occupied permanently by people.

Contemporaneously, the wars and the Seminole's escape into the swamps without surrender brought American adventurers into the area, who found the muck lands of the Everglades to hold intriguing promise for agriculture. Drainage of lands for agriculture was firm government policy from 1850 and the resulting story of the drainage of the Everglades is well known.[16] By 1881, 9,800,000 ha of the Everglades had been sold by the state of Florida for drainage, and, through various incarnations, the process continued for 40 years. Drainage was a societal imperative that took several successive approaches as the decades passed by, each driven by the social pressures of the day leading eventually to the management of undrainable land for flood control and water management.[17] Drainage was also, of course, driven by society's new technological abilities. The two dominant pressures benefiting from drainage were settlement, which increased during the period, and agriculture, which eventually took hold on drained land south of Lake Okeechobee and on seasonal lands along the edges of the inland marsh.

The settlement of south Florida by Americans was not an easy task, but one that was well underway within a decade of the appearance of the first dredges. Although most of America's pioneering spirit was directed to the west of the continent, a few saw the opportunities of south Florida. Settlement of American pioneers along the east coast began at the same time as drainage, starting on the high ground along the coasts. Market hunters efficiently penetrated well inland, using firearms and traps to hunt alligators, raccoons, otter, and plume birds; populations of each in turn rapidly crashed. For example, commercial fishing began on Lake Okeechobee in 1883, as soon as the dredges had worked a couple of years. After the American Civil War (1870), there were 85 residents of eastern south Florida, derived from the American north, Bahamas, and Europe.[18] These people built rough houses, hunted, fished, gathered, farmed, and harvested coontie (*Zamia pumila*, a cycad that the Indians had taught pioneers to make into a marketable flour). Their settlements occupied the same high grounds as the ancient Calusa, Jobe, and Tequesta villages. The area was visited by wealthy yachtsmen and sportsmen, who hunted big game in the pinelands and large fish along the coast. These American settlers differed from their predecessors in their incessant attempts to farm the rocky and wet land, soon aided by the machinery of the industrial revolution. Within 20 years these new residents were able to use technological innovations to dig through the muck and rock, build extensive canals, elevate land, and fill bay bottoms in ways that had been impossible before. The turning point of the American period was the enticement of Flagler's railroad to the small town of Miami in 1896.[19] By that time people had again settled all along the coast, but the railroad created a city in Miami, unbeknownst to them the site of one of the Tequesta Indian villages. The history from that moment to present is one of ever-increasing population growth, colonisation of recently drained land further from the coast, and modernisation, with the attendant social pressures and evolution.[20] This evolution of the human development of the Everglades is an example of Lenski's[21] contention that it is the technology of a given society that spurs cultural change. This is demonstrated in historical man-made environmental changes that took place (and paralleled technological developments) from early hunter-gatherers in Florida to the rapid development that occurred in the 20th century with the onset of improved technology.

## The Current Culture of South Florida

The current cultural heritage of south Florida is a study in contrasts. On the increasingly unflooded land of south Florida, mega-cities have proliferated over the past 120 years. Florida's population has been doubling every 20 years, resulting in a 2000 census of 6,290,000 residents in the four counties of south Florida.[22] The modern inhabitants have highly diverse cultural backgrounds. Racially, 70% of the population is white, 20% black, 1.7% Asian and, importantly to our history, 0.2% remains American Indian.[23] By ethnic origin, 40% of the population of south Florida is Hispanic, which in itself is a far from homogeneous group. The dominant group is Cuban (18% of the population), but 21% come from other nations including Puerto Rico (3.4%), Mexico (1.5%), as well as sizeable communities from Nicaragua, the Dominican Republic, Peru, Honduras, Venezuela, Colombia, and Brazil.[24] Non-Hispanic Jamaicans and Haitians also have substantial communities. In 2000, over a half million south Floridians identified themselves as 'some other race' from those on the standard list being offered—Mayans and Haitians, for example, seem to disavow existing narrow ethnic categories.[25] From 1990 to 2000, the increase in foreign-born residents in south Florida was nearly half a million people.[26] Among the fastest growing are immigrants from Haiti. They come from the poorest nation in the Western Hemisphere, have limited resources and education, speak a very different language, and find themselves without special benefits, occupying the bottom of the economic scale, although they too are rising economically.[27] Despite the relatively rapid economic move of immigrants to Miami into the American middle class, the economic extremes in south Florida are indeed extreme. The median household income of $35,966 is one of the lowest of all major cities in the USA, and in Miami-Dade County, 2.7% of households make over $200,000 per year but 15% of households are below the poverty line.[28]

Ethnic diversity has led to economic diversity and political strength. All ethnic groups are active in, and indeed some dominate, the local economy. Shopkeepers profit from the diverse demands of different nationalities. Much of the normal business practice in south Florida is now owned and operated by Hispanics, particularly Cubans who now have had two generations to establish themselves. Added to the usual business culture are people from many countries working in national and international industries centred in Miami—international finance, import, law, shipping, pleasure-cruising and other tourism-based activities. With population and economic strength comes political strength. Politically the Cuban community of south Florida has a favoured standing with the government of the USA, a standing strong enough to wield influence at national elections and in international relations.

As in economics and politics, culturally Miami and south Florida today are a matrix of divergent yet intermingling influences. While traditional American customs are followed in south Florida, each nationality also brings its own customs and values. Added to the American holidays are festivals, participated in by all segments of the community, highlighting cultures such as Cuba (Calle Ocho Festival), the Bahamas (Junkaroo), Mexico (Cinco de Mayo), and Haiti (Haitian Independence Festival). Added to the traditional American religions are Santeria from Cuba and Voodoo from Haiti.[29] Woven into the city landscape are ethnic enclaves such as Little Havana and Little Haiti. Spanish, not English, is the language expected in much of economic intercourse. South Florida today is both rich and poor; it is Anglo, African American, Hispanic and Creole; it has families that are economically depressed and businesses that are international in scope and power; it is politically balkanised yet influential nationally and internationally.

A universality of south Florida is that all its inhabitants, including Native Americans, may be considered to be relatively recent immigrants. Many of these arrived in Florida from international places of origin. However, those from all places north of southern Florida (yet still in the USA) can also be considered 'intra-national' immigrants themselves. They have, after all, migrated within their own country to a vastly different sub-tropical climate filled with plant and animal species, soil and landscape that is distinctly different from most of the USA. In addition, Florida post-1900 has had an international flavour, which began with early Bahamian settlers and continues today with the prevalence of foreign languages, global arts, and exotic foods. Therefore, US northerners have also been cultural immigrants. As we examine these immigrant groups in south Florida, it should be asked whether the status of 'immigrant' has affected the way in which arrivals to Florida have interacted with the natural environment during the past century. Since so many people in south Florida moved here from elsewhere, has this affected their relationship to place and the way in which they perceive and/or exploit natural resources? There are two aspects: the cultural traditions from which the immigrant derived and the persistent sense of impermanence.

A common feature of most of the cultural heritages of the immigrants to south Florida is that they come from cultures that have limited experience with environmental protection and/or are highly exploitative of their natural environment. Cuba, Haiti, Jamaica, and Central America were, and largely remain, quite rural and agricultural and poor, a legacy of their colonial past. Such immigrants bring with them little environmental consciousness and tend to perceive nature as infinite or having little value.[30] In other sections of the USA, such as California, evidence suggests that immigrants have played a role in burgeoning environmental justice issues.[31] And Pfeffer and Stycos have found that stereotypes of immigrants as less concerned about the environment in New York City were unsupported, although indeed they were less likely to participate in the 'mainstream' pro-environment movement.[32] Blacks from Haiti come from a country that is one of the most extreme examples of a natural heritage destroyed. Upon arrival in south Florida, most Haitians have little personal experience in environmental protection. Intra-national black immigrants to south Florida derive from settings with little experience in environmental conservation and confront overwhelming social issues. Recent research suggests that black intra-national immigrants to Miami who grew up with either agricultural roots in the southern USA or in the Bahamian culture may have developed connections to nature and perhaps enhanced environmental concerns.[33] And Haiti's situation may not indicate that its people value the natural environment any less; instead, simply that they value survival. How immigrant values and experiences may translate into a new cultural setting in south Florida remain to be seen. A background of poverty and political displacement does not predispose individuals to care much about the natural heritage of a new land, especially a culturally unfamiliar land to which they recently migrated.

International immigrants from Cuba, Haiti and other distant lands may share an additional and important characteristic with the intra-national US immigrants that may matter more than their land of origin. The majority of individuals in both groups considered themselves 'short-timers' upon their arrival in Florida. Early Cuban immigrants often believed that they would return home and saw south Florida as a place of temporary asylum. The same may be said of some Haitians, of Venezuelans and Colombians fleeing political problems, economic issues, and violence. Many in the first generation hope that situations in their country of origin will improve so that they might return. Intra-national immigrants from US cities such as New York, Boston and Chicago were often retirees with few relatives and limited future stakes in Florida. Other intra-nationals were part-time residents, people buying second homes, temporary labourers, 'get-rich-quick' schemers, people lured by the development boom seeking economic incentive, and people searching for temporary respite from colder climates none of whom envisage south Florida as 'home'. Unlike the ancient Calusa Indians, later people arrived with the option, and sometimes the hope, of leaving if the boom went bust, if the land were despoiled, if the aquifers dried up or if their native land from Nicaragua to New York beckoned their return. Short-term investments and profits and temporal 'place' commitments appeared to lead to short-term local environmental polices and either aggressively abusive or laissez-faire attitudes towards natural resources.

The international and intra-national transport of these views of impermanence and resource undervaluation to south Florida in the past several decades has increased pressures for the development of the remaining wetlands in south Florida. In the past two decades, the federal and state governments have taken dramatic measures aimed at preserving or restoring the Everglades.[34] Yet in this same period, development has pushed to the very edge of the levees surrounding the remaining natural Everglades, imperilling water supplies and even occasioning the state of Florida to constrain local development proposals.[35] There seems, in fact, to be a fundamental policy gap between national and local aspirations with respect to conservation of the local natural heritage.

## Persistent Patterns of Human Culture in South Florida

From the earliest times of human occupancy in south Florida, home sites were concentrated on high ground. Sites were elevated to the greatest extent technologically possible using available materials and manpower. Where possible, canals were dug, firstly by hand. Levees and seawalls were built with mounds of shell. An economy based on fishing, minor hunting, gathering and trade developed and persisted. With the exception of the Spanish, who for the most part engaged only marginally in Florida, and the Seminoles, who were forced by ethnic cleansing into the previously uninhabitable Everglades, the methods of occupying south Florida did not change fundamentally from the development of the Everglades, 5,000 years BP, until the late 19th century AD—from the Archaic, through the Calusa, to the American periods. And throughout this time human occupancy of south Florida was inherently exploitative. From prehistory to the industrial revolution, however, exploitation was limited to a subsistence-level existence for a population in the thousands managed by locally adapted cultural traditions.

What changed, starting in the late 19th century, was the employment of mechanised means to achieve a national vision that favoured settlement and agriculture of this seemingly open land. Prior to this time, large-scale settlement, full-time occupancy of the interior Everglades, extensive drainage, and agriculture were never part of the interaction of man and environment in south Florida. But the evidence suggests that it was not for lack of desire or even of effort, but rather a matter of not being able to accomplish this goal with available technology. So the modern drive for settlement, attendant population increases and progressive encroachment into remaining seasonal wetlands may be seen as a continuation of the processes of thousands of years of human occupancy of south Florida. The environmental and political dominance of agriculture is, indeed, a new feature; one that is a driving force in contemporary environmental degradation.[36] Another force is pressure for living space, given that in less than 120 years the human population increased from fewer than 100 to over 6 million people. Interestingly, exploiting marine animals continues to be an important part of the local culture. In Florida as a whole over 103 million pounds of seafood are taken annually[37]—a modern reflection of the marine economy that sustained the Archaic Indians of south Florida. However, the queen conch, which supplied the high ground for home sites for generations of Indians, is now a threatened species.

National policy now requires adherence to environmental standards. Much of the remnant wetland is under the control of national and state governments. While national regulations control the drainage and infilling of wetlands and the quality of surface water, these remain political, as well as environmental, issues. There have been recent rollbacks of some standards and laws that provide environmental protection, including issues over the definition and protection of wetlands. And the issue remains that local governments control local development. The apparent desire for continued development exhibited by the highly diverse local population, most of whom are first- or second-generation international or intra-national immigrants to south Florida, may be in some small measure balanced by national and state policy. Of interest is that with the legal and publicity wars over the south Florida environment, the remaining Indian nations of south Florida (who have recently gained an economic advantage through their support of gaming and other activities not allowed in the rest of Florida), have become leaders at the forefront of efforts to protect the Everglades from future development. Is it that the Indian nations of south Florida have the deepest sense of permanence within the environment? Deep societal tensions remain between exploitation and conservation. Of course, the tension is unbalanced since most of the human-induced changes in the natural cultural heritage are irreversible, whereas the local human culture can choose to change.

## Conclusions

The interaction between people and the environment in and around the Everglades of southern Florida is a story of international renown. From its beginning, even as the Everglades and local culture co-evolved, this interaction has been inherently exploitative of the environment. It was, for most of this period, also one of respect for that environment driven by seasonal uncertainty in the availability of critical resources, which deeply influenced religious, economic, and political practice. The respect for the natural heritage shown by indigenous Calusa was born of the fact that they could not control the annual instabilities other than ritualistically. The result was a society that existed sustainably because population size and dispersion, political power and social complexity were limited by those resources. Early Indian society in south Florida was ecologically sustainable because it had to be. This culture made little or no impact on the functioning of the natural environment, because it could not.

The 150-year hiatus in man's occupancy of south Florida provided an interlude, perhaps unique in North America, in which the natural heritage proceeded unimpeded in any way by humans, a period allowing insight into how the landscape would be functioning today without the influence of people. If prior to this hiatus human impact was minimal and population capped at sustainable levels, after the hiatus the new culture's ability to manipulate the landscape led to the imperative to do so and the exploding population has not yet seen its new cap. The result of the last 120 years is that much of the natural heritage of south Florida has been destroyed. It remains to be seen whether the adverse effects of degradation of the natural environment and/or the potential social or political pressure of a strong environmental ethic (whether originating at a grassroots local level, from the state, from the Indian nations, or from the national government) will inspire or provoke change towards a more sustainable environmental ethic. Could one or more of these scenarios eventually encourage a more respectful interaction with the remaining natural environment among a transplanted local population whose own heritage, from California to Cuba, from Hawaii to Haiti, has been environmentally exploitative?

Internationally, nationally and regionally, protection of the remnant Everglades environment is accepted as good public policy. But the details of what is good environmental policy for the remaining Everglades remain highly politicised and hotly debated. The approved programme to enhance flood control, security of water supply, and preserve part of the Everglades enjoys a governmental commitment of $130 billion over 30 years, yet only a small proportion of these funds is scheduled to be spent on actual environmental restoration.[38] The adverse effect of environmental degradation on the economy, social viability, and sustainability is well understood, even in south Florida.[39] Locally, however, the desire for, or at least the tolerance of, environmental exploitation continues unabated among the majority of the newly arrived population, whether they hail from Havana, New York, Caracas, or Quebec. The emergence of the newly wealthy resident Native American tribes as leaders in environmental protection may be the most fundamental example of the role of permanency. Admittedly, south Florida could not be home to 6.3 million people without intense management of the seasonal fluctuations that also deeply affected the indigenous cultures. The question now is—what is the human carrying capacity of this new environment? In the past 5,000 years, human populations have fluctuated in concert with climate, resources, and socio-political adaptations. What population can south Florida now sustain, how can it be supported, and at what stage does it overwhelm the ability of the local environment? What point is the point of environmental 'no return'? Will the local community adopt the ethic of living within the constraints of the local environment, as is being urged by some in the larger environmental and political community? Can a local adaptation of the present cultural practices of the people of south Florida occur? Changes do come rapidly to immigrant peoples, especially in the third generation through education and personal economic advancement. Perhaps most importantly, will immigrants overcome an aura of impermanence and arrive at a vision of south Florida as embodying a permanent sense of place? It is easier to exploit a land and its resources when it is only a stopping point, a retirement layover, the location of a holiday home, a city you go to for a promotion, or a destination to wait in until your own country becomes politically or economically stable. If the land of south Florida becomes 'home', immigrants become stakeholders and may be convinced that short-term benefits inherent in environmental exploitation have pervasive long-term costs.

As shown in this paper, filling and elevating land, canal digging and resource exploitation have always been a dominant feature of human interaction with the Everglades, since before there was an Everglades. Viewing the human story as a continuous one puts it in a different, and perhaps more helpful, perspective than the more common view of environmentally respectful Indians and environmentally exploitative moderns. Both were attempting to live, eat and organise socially in an inherently inhospitable environment. Until the contact period, people had adapted their culture to sustain communities in balance with this difficult landscape. After the late 1880s, humans changed the equation and managed to do what ancient peoples could not, namely drastically alter the landscape itself with the assistance of modern technology. Man eventually won this interaction, even if perhaps only for the short term.

## Notes

[1]  Kushlan, 'Freshwater Marshes', 329.
[2]  Zeiller, *A Prehistory of South Florida*, 29; Hoffmeister, *Land from the Sea*.
[3]  Zeiller, *A Prehistory of South Florida*, 50.
[4]  McCally, *The Everglades*, 31.
[5]  Ibid.
[6]  Ibid., 31–57; Tabeau, *Man in the Everglades*, 37.
[7]  Zeiller, *A Prehistory of South Florida*, 93–110; McCally, *The Everglades*, 47; Gilliland, *The Material Culture of Key Marco*.

[8]    Milanich, *Archeology of Pre-Columbian Florida*, 277; Zeiller, *A Prehistory of South Florida*, 74–149; McGoun, *Prehistoric Peoples of South Florida*, 60–64; McCally, *The Everglades*, 74; Marquardt and Payne, *Culture and Environment in the Domain of the Calusa*.

[9]    Fontenada, *Memoir of Do d'Escalente Fontaneda Respecting Florida*, 11–20.

[10]    McCally, *The Everglades*, 46.

[11]    Widmer, *The Evolution of the Calusa*, 260.

[12]    McCally, *The Everglades*, 43.

[13]    Ibid., 54.

[14]    Ibid.; Henderson and Mormino, *Spanish Pathways in Florida*.

[15]    Covington, *The Seminoles of Florida*.

[16]    Blake, *Land into Water–Water into Land*.

[17]    Ibid.; Solecki et al., 'Human–Environment Interactions in South Florida's Everglades Region'.

[18]    Peters, *Biscayne Country*, 6.

[19]    Standiford, *Last Train to Paradise*.

[20]    Allman, *Miami*; McCally, *The Everglades*; Tabeau and Marina, *A History of Florida*; Grunwald, *The Swamp*.

[21]    Lenski, *Power and Privilege*; Nolan and Lenski, *Human Societies*.

[22]    South Florida Regional Planning Council, South Florida Census 2000 Resource Center, United States Census 2000 [accessed 15 January 2007], available from www.sfrpc.com/census/mapssofla1.htm, 2006.

[23]    South Florida Regional Planning Council, South Florida Census 2000 Resource Center, United States Census 2000 [accessed 15 January 2007], available from www.sfrpc.com/ftp/pub/census/SFCPL2k3.PDF, 2006.

[24]    Portes and Mozo, 'The Political Adaptation Process of Cubans and Other Ethnic Minorities in the United States'; Garcia, *Havana USA*; Florida Regional Planning Council, South Florida Census 2000 Resource Center, United States Census 2000 [accessed 15 January 2007], available from www.sfrpc.com/ftp/pub/census/gdc2k_sfl.pdf, 2006.

[25]    Mormino, *Land of Sunshine, State of Dreams*, 295.

[26]    South Florida Regional Planning Council, Regional Profile [accessed 15 January 2007], available from http://www.sfrpc.com/ftp/pub/srpp/regional%20profile.pdf

[27]    Sohmer et al., *The Haitian Community in Miami-Dade* [accessed 15 January 2007], available from http://www.brook.edu/metro/pubs/20050901_haiti.htm

[28]    South Florida Regional Planning Council, South Florida Census 2000 Resource Center, United States Census 2000 [accessed 15 January 2007], available from www.sfrpc.com/ftp/pub/census/SF3MiamiDade.pdf, 2006.

[29]    Canizares, *Walking with the Night*; Hurbon, *Voodoo*.

[30]    Malonado, 'Cuba's Environment'.

[31]    G. Flaccus, 'Latino Immigrants Embracing Green', *The Monterey Herald*, 16 October 2006 [accessed 18 October 2006], available from http://www.montereyherald.com/mld/montereyherald/news/state/15770658

[32]    Pfeffer and Stycos, 'Immigrant Environmental Behaviors in New York'.

[33]    Smith-Cavros, 'Black Churchgoers'.

[34]    Grunwald, *The Swamp*; Comprehensive Everglades Restoration Plan [accessed 15 January 2007], available from www.evergladesplan.org, 2006.

[35]    Ibid.; 'State Warns Miami-Dade against New Projects', *South Florida Business Report* [accessed 15 January 2007], available from http://southflorida.bizjournals.com/southflorida/stories/2006/07/03/

[36]    Wilkinson, *Big Sugar*.

[37]    Florida Department of Agriculture and Consumer Affairs, 'Overview of Florida Seafood and Aquaculture' [accessed 15 January 2007], available from www.fl-seafood.com/industry/agfacts.htm, 2006.

[38]   Comprehensive Everglades Restoration [accessed 15 January 2007], available from www.ever-gladesplan.org; Grunwald, *The Swamp*.

[39]   Munasinghe, *Environmental Economics and Sustainable Development*; E. Moncarz, A. Jorge and R. Moncarz, 'The PreCuban Miami Economy', paper presented at the International Trade and Finance Association 15th International Conference, paper no. 24, 2003.

# References

Allman, T. D. *Miami: City of the Future*. New York: Atlantic Monthly Press, 1987.

Blake, N. M. *Land into Water–Water into Land: A History of Water Management in Florida*. Gainesville: University Presses of Florida, 1980.

Canizares, R. *Walking with the Night: The Afro-Cuban World of Santeria*. Rochester, VT: Destiny Books, 1993.

Covington, J. W. *The Seminoles of Florida*. Gainesville: University Presses of Florida, 1993.

Fontenada, H. D. *Memoir of Do d'Escalente Fontaneda Respecting Florida*. Translated by Buckingham Smith. Edited by David O. True. Miami: University of Miami Press. First published 1854; reprinted 1944.

Garcia, M. C. *Havana USA: Cuban Exiles and Cuban Americans in South Florida, 1959–1994*. Berkeley: University of California Press, 1997.

Gilliland, M. *The Material Culture of Key Marco, Florida*. Gainesville: University Presses of Florida, 1989.

Grunwald, M. *The Swamp: The Everglades, Florida, and the Politics of Paradise*. New York: Simon & Schuster, 2006.

Henderson, A. L. and G. R. Mormino, eds. *Spanish Pathways in Florida, 1492–1992*. Sarasota, FL: Pineapple Press, 1991.

Hoffmeister, J. E. *Land from the Sea: The Geological Story of South Florida*. Coral Gables: University of Miami Press, 1974.

Hurbon, L. *Voodoo: Search for the Spirit*. New York: Abrams, 2002.

Kushlan, J. A. 'Freshwater Marshes'. In *Ecosystems of Florida*, edited by Ronald Myers and John J. Ewel. Orlando: University of Central Florida Press, 1990.

Lenski, G. *Power and Privilege: A Theory of Stratification*. New York: McGraw-Hill. 1966.

Malonado, A. 'Cuba's Environment: Today and Tomorrow—An Action Plan'. *Cuba in Transition* 13 (2003): 63–73.

Marquardt, W. H. and C. Payne, eds. *Culture and Environment in the Domain of the Calusa*. Gainesville: Institute of Archeology and Paleoenvironmental Studies, University of Florida, 1992.

McCally, D. *The Everglades: An Environmental History*. Gainesville: University Presses of Florida, 1999.

McGoun, W. E. *Prehistoric Peoples of South Florida*. Tuscaloosa: University of Alabama Press, 1993.

Milanich, J. T. *Archeology of Pre-Columbian Florida*. Gainesville: University Presses of Florida, 1994.

Mormino, G. R. *Land of Sunshine, State of Dreams: A Social History of Modern Florida*. Gainesville: University Presses of Florida, 2005.

Munasinghe, M. *Environmental Economics and Sustainable Development*. Environment Paper no. 3. Washington, DC: World Bank, 1993.

Nolan, P. and G. Lenski. *Human Societies: An Introduction to Macrosociology*. 9th ed. Boulder, CO: Paradigm, 2004.

Peters, T. *Biscayne Country, 1870–1926*. Miami: Banyan Books, 1981.

Pfeffer, M. and J. Stycos. 'Immigrant Environmental Behaviors in New York City'. *Social Science Quarterly* 83, no. 1 (2002): 64–81.

Portes, A. and R. Mozo. 'The Political Adaptation Process of Cubans and Other Ethnic Minorities in the United States: A Preliminary Analysis'. *International Migration Review* 19 (1985): 35–63.

Smith-Cavros, E. 'Black Churchgoers, Environmental Activism and the Preservation of Nature in Miami, Florida'. *Journal of Ecological Anthropology* 10, (2006): 33–44.

Sohmer, R. R., D. Jackson, B. Katz and D. Warren. *The Haitian Community in Miami-Dade.* Washington, DC: Brookings Institution Metropolitan Policy Program.

Solecki, W. D., J. Long, C. C. Harwell, V. Myers, E. Zubrow, T. Ankersen, C. Deren, C. Feanny, R. Hamann, L. Hornung, C. Murphy and G. Snyder. 'Human–Environment Interactions in South Florida's Everglades Region: Systems of Ecological Degradation and Restoration'. *Urban Ecosystems* 3 (1999): 305–43.

Standiford, L. *Last Train to Paradise: Henry Flagler and the Spectacular Rise and Fall of the Railroad that Crossed an Ocean.* New York: Crown, 2002.

Tabeau, C. W. *Man in the Everglades.* Coral Gables: University of Miami Press, 1968.

Tabeau, C. W. and W. Marina. *A History of Florida.* 3rd ed. Coral Gables: University of Miami Press, 1999.

Widmer, R. J. *The Evolution of the Calusa: A Nonagricultural Chiefdom on the Southwest Florida Coast.* Tuscaloosa: University of Alabama Press, 1988.

Wilkinson, A. *Big Sugar.* New York: Alfred A. Knopf, 1989.

Zeiller, W. *A Prehistory of South Florida.* Jefferson, NC: McFarland, 2005.

# Natural World Heritage: A New Approach to Integrate Research and Management

Jean-Claude Lefeuvre

## Evolution of Natural and Modified Systems

The Quaternary era was marked by an ever-changing climate, characterised by strong variations in temperature; glaciations alternated with interglacial periods for at least 1 million years. In parallel, the sea level varied considerably. The amplitude of these changes of level could exceed 120 m. In this context our species, *Homo sapiens*, knew how to conquer new territories and how to continue its evolution. On this subject de Lumley has written: 'endowed with extraordinary capacities of adaptation, the Human being will develop his civilisations in all latitudes, in all climates, in all landscapes and will survive in extreme conditions. When his environmental factors become most difficult his cultural leaps will be the largest.'[1] In fact, it was in the middle of the Ice Age, more than 35,000 years ago, that parietal art developed, leaving us caves decorated with admirable representations of vertebrate species, both as hunted animal and troublesome predator.

Actually, it is from this time and until approximately 9,000 years ago that one can witness a genuine first great revolution which makes inseparable the expression of cultural phenomena marking embryonic civilisations and new ways of living in ecosystems increasingly modified by human activities and subjected to human needs. This revolution resulted in the transition from a nomadic to a sedentary lifestyle, with the construction of huts made from branches erected above bases of gathered stones (the first villages), and worship of the dead. Later, in the context of climatic change and in an area marked by an abundance of natural resources, plants and animals available for gathering and hunting, came the beginning of selective intervention in the ecosystem with the grading of coarse grain cereals, flour making and bread baking. This was in the Middle East, part of which was to be called the 'fertile crescent'. These changes were a prelude to the birth of agriculture, with the selection of wild fig trees, mutant, sterile 11,400 years ago in the Jordan valley,[2] followed by the cultivation of cereals and the domestication of wild ungulates, accompanied by the building of the first villages with dwellings and terracotta silos.

The Neolithic revolution was launched. It was to extend on both sides of the Mediterranean Sea and all over Europe in less than 5,000 years.[3] Little by little the natural ecosystems were profoundly modified throughout Europe for agricultural needs as well as for stone from the Earth's crust for use in the building of permanent habitats. One can say that from this period onwards, all human construction, all the monuments which attest to our cultural evolution, fit into nature as adapted by human beings, directed towards the production of human well-being and services. Admittedly, when we refer to one of the civilisations which marked the history of humanity—that of the Egyptians—we can still say that nature spoke to the Egyptians, creators of an exceptional monumental art, amazing frescos and bas-reliefs[4] on which nature is present everywhere, in particular in the form of an enduring reference to two plants from the wetlands, the papyrus, king of the North, and the lotus, the plant of rebirth. And what can we say about the representations of flights of ducks, Egyptian geese or fish captured with a net, if not that 'the universal message of the civilisation built by Egypt lies in the respect shown for the world and the forces which govern it',[5] like the sun god Râ represented by a man with a falcon head surmounted with a solar disc or in the form of a crowned beetle. The gods of Egypt claim to be part of various wild species of fauna, crocodile, wild dog, falcon, lioness. But in parallel, the reference to wild nature is altered by the reference to nature as modified for the breeding needs of the ram in particular and also of the crowned cow Hator which holds an important place in the Egyptian polytheist religion. One should not forget either that the wetlands lining the sacred River Nile were controlled and modified to become cultivated fields using the annual floods which permeated Egyptian life.

The extension of the 'Neolithic revolution' all across Europe and the growth in the number of human beings on European territory in various civilisations—from the East (e.g. Celts) or South (Roman Empire)—were accompanied by one period of regression or extraordinary degradation of the natural ecosystems, the forest in particular, which was to continue for centuries. Cereals need open spaces, requiring the creation of glades by clearing. In addition to the development of agriculture, the proto-industry of iron or glass, the transformation of limestone into lime, the requirements of structural timber for the cities or the shipbuilding industry, the demand for charcoal in the cities, all contributed to the regression of forests. Thus, in France, the surface covered by the forest decreased from 44 million ha in 5,000 BP to 7.7 million ha just before the French Revolution in 1789. In parallel, what was left of the forest was subjected to the right of pasturage, acorn gleaning or to the felling of trees selected according to need (oak for the saw log, shipbuilding or tannin, beech for clogs and the wood craft industry) and where structure depended on its use (from clusters with large trees for the saw log to clumps used for firewood). All these spaces could no longer be compared to natural environments after so many human interventions. Little by little the large herbivores such as aurochs, bison and the large predator populations (bear, lynx, wolf) decreased. This evolution was carried out in spurts as a result of invasions, wars, epidemics, famines and climatic changes (e.g. the 'little ice age') with an alternation of fallow lands (e.g. during the One Hundred Years War) and clearing lands.

According to de Ravignan et al. this evolution resulted in a European territory characterised by rural landscapes where three generally complementary parts could be distinguished: a cultivated space or *ager*, a pastoral space or *saltus* and a forest space or *silva*.[6] It is the layout of the *ager* that determines the dominating aspect of the rural landscape. This is much diversified, as much at the European level as on French territory: from the Beauceron open field to the Breton *bocage* or the Cevennes terrace fields (see Figure 1). Agricultural systems based on mixed farming and mixed breeding, adapted to the soil and the climate, strongly influenced the functioning of these various landscapes to which flora and wild faunas adapted, as described in Clavreul (see Figure 2).[7] Some of these systems became highly artificial, such as the *bocages* with their agricultural parcels surrounded by ditches and slopes covered with quickset hedges and periodically pollarded trees (generally every nine years). These systems could have a more important biological diversity than the systems qualified as 'natural' and located next to them.[8] The modernisation of agriculture characterised by the 'improvement' of plants and domestic animals, increasing mechanisation, the use of chemicals (fertilisers, pesticides, etc.), the extension of single-crop farming and single breeding, industrial breeding, etc. completely changed this vision again. The plant and animal species had difficulty in adapting to these new conditions. This is demonstrated in Clavreul's study (see Figure 2) of the various landscapes in the same territory which result from such a change,[9] and is what Binder (see Figure 3) also confirms.[10] The

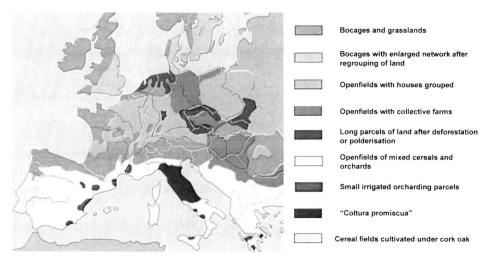

Bocages and grasslands

Bocages with enlarged network after regrouping of land

Openfields with houses grouped

Openfields with collective farms

Long parcels of land after deforestation or polderisation

Openfields of mixed cereals and orchards

Small irrigated orcharding parcels

"Coltura promiscua"

Cereal fields cultivated under cork oak

**Figure 1**    Rural landscapes in Europe at the middle of the twentieth century. (From Brunet in De Ravignan *et al.*, 1990).

evolution of bird settlements in various types of habitats (see Figure 4) reinforces the idea that the transformation of agricultural territory negatively impacts biodiversity.[11]

Thus, humans transformed not only most European territory in 3,000 years of history by substituting modified agro-systems for the land's natural ecosystems or by abandoning some of these modified ecosystems after use, but also by controlling the rivers with dykes, and transforming the functioning of the aquatic, continental or marine systems by unfavourable nutrient enrichments (eutrophication) as with, for example, the compounds derived from inorganic or chemical synthesis. We thus now live in an eminently changing environment that contrasts strongly with the supposedly immutable, timeless character of many monuments registered on the World Heritage List. Admittedly, these cultural traces of various civilisations which mark the human history of our planet have suffered enough from human activities. Indeed, Fabre shows that the bridge-aqueduct of Gard (registered in 1989) built by the Romans to supply water to the town of Nimes and which remarkably has resisted the ravages of time and floods, has suffered from loss of material (stones were removed for Lidenon Castle and for the church of St-Bonnet-du-Gard).[12] The left bank abutment was cut down by 130 m from its original height and the bridge-aqueduct also lost its breast wall. Another example is Angkor, the monumental and archaeological site of Kampuchea, revealed to the world in 1860, left in a state of abandon, and often plundered from 1972, before its registration as a World Heritage Site in 1992, and which now requires important restoration work. We also know that the air pollution that we generate harms the original stonework of some of these precious monuments. It has been possible, however, to find the quarry which provided the original stones and carry out repairs in order to preserve these outstanding elements of human history for future generations. But with the evolution that we have imposed on nature, the disappearance or regression of

Richness (number of species)

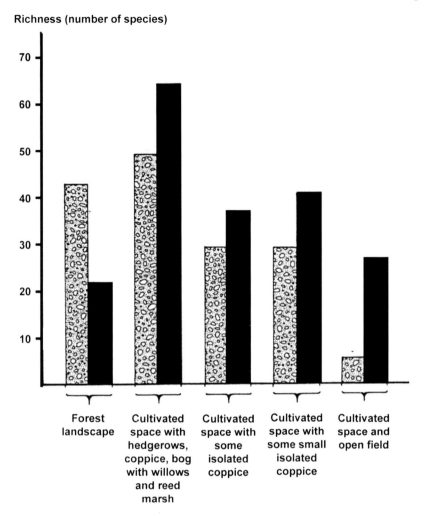

**Figure 2**   Richness of birds (grey bar) and carabid beetles (black bar) communities in various rural landscapes. (From Clavreul, 1984.)

certain species, the invasion by foreign species, and the modification of ecosystems as a result of our activities, what strategy can we develop to preserve the future of the 'natural monuments' in accordance with Alexander Von Humboldt's meaning, knowing that the 'bricks of life', the species, are irreplaceable, at least on a temporal scale, for thousands of human generations?

### Mont Saint-Michel and its Bay: A Standard of Association between Nature and Culture

The first French site registered on the UNESCO World Heritage List for both cultural and natural reasons, Mont Saint-Michel and its bay do indeed seem to lend themselves to a comparative analysis between what is done to protect a prestigious monument

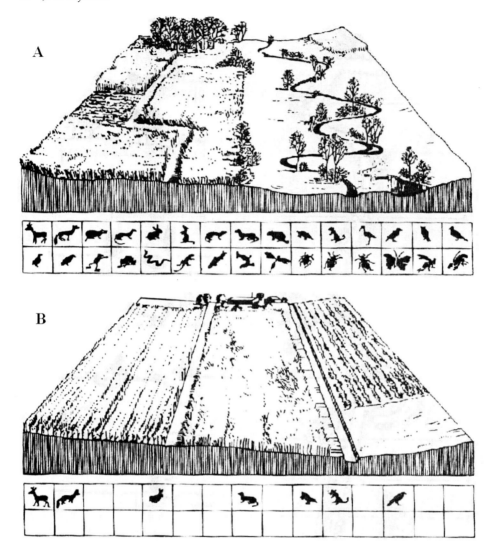

**Figure 3**   Changes in rural landscape from mixed farming (A) to single crop farming (open field) (B). (From Binder, 1986.)

anchored firmly to a rock, the Mont Tombe, known worldwide, visited annually by more than 3 million visitors, and what could be done to preserve the casket of this jewel (the bay) sometimes referred to as the Wonder of the Occident.[13] The mount was a monastery from 966; then in 1790 it lost its abbey status to become a state prison, a status which it held until 1863. In 1867, monks returned to the monastery and in 1872, with the appointment of the architect Corroyer, thoughts turned to repairing the damage caused to the buildings; at last the time for restoration had started. The spire surmounting the abbey church was built and then capped with Saint

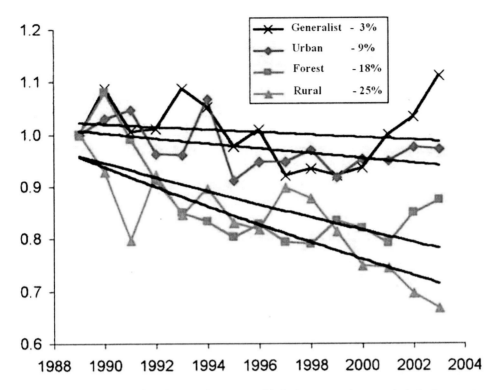

**Figure 4**   Change in abundance of common birds in France (95 species) during 1988–2004, in rural landscapes, forests and urban areas. The French wild bird indicator is based on the population trends of wild breeding birds. (From Couvet *et al.*, 2004.)

Michel's statue (by Frémiet, 1897). The restoration work currently continues in the context of a major project to re-establish a maritime character around Mont-Saint-Michel. This development started in February 2006 and serves to remind us that the bay has been subjected to changes since its origin. In fact, it represents a remarkable model of a littoral area, originating at the time of the Flandrian transgression and largely reorganised by human intervention over a period of about 15 centuries. Mont-Saint-Michel bay is nearly 50,000 ha in extent, distributed between a maritime domain (approximately 30,000 ha of which more than two-thirds are above normal tides), old and recent polders (15,000 ha), and valleys and continental wetlands (approximately 3,000 ha). Its entire catchment area approaches 3,250 km$^2$. It is a beautiful example of the transformation wrought by nature over the centuries. Indeed, in this bay, formed 6,000 years ago in its current configuration, under the action of a 'natural' rise of sea level, due to the climatic re-warming following the last glaciations (and which created the English Channel, after a sea-level rise of more than 120 m), the first sedimentary settlements that clogged the bottom of the bay were already being more or less modified, from 3,000 years ago, by the arrival of terrestrial elements coming from the first clearings. Then, one part of the saltmarshes was cut off from the sea by a shelly reef, the 'natural' increase of which could have been accentuated by the shells' use for food, of which the oysters (*Ostrea*) abundant in the bay form a large part. Certainly, from the 11th century onwards this shelly reef was used as a support for a dyke built by the dukes of Brittany. It was completed more than two centuries later, isolating the first polders of the bay, known as 'Marais de Dol'. These 12,000 ha of saltmarshes dominated by halophilous plants were thus to be transformed into soils for crops and breeding and were quickly dominated by terrestrial plants. On the other hand, those parts that were difficult to drain were to be populated by wetland plants characteristic of freshwater environments.[14]

After the attempts to settle the polders between Mont-Saint-Michel and the mainland had failed (the Quinette de la Hogue concession), the French state was to concede a vast territory on both sides of Mont-Saint-Michel to the Dutch company Mosselman & Donon. This company, after the Couesnon canalisation in 1856, accomplished the settlement of polders in the western part while gradually isolating, by 1935, 2,200 ha of arable land by successive dams. These became known as the 'recent polders'. Only a few polders were to be set up in the eastern part, and the Roche Torin dyke, planned to connect this locality to Mont-Saint-Michel, was never to be finished due to storm damage and the movements of the Rivers Sée and Sélune. All these developments, the construction of a dyke-causeway and a relatively recent estuary dam on Couesnon, seem to be responsible for most of the sediment concentrates now in the estuary part of the bay, near Mont-Saint-Michel. These sedimentary settlements, which in some places can reach over 3 cm per year, allow the progression of saltmarshes in front of the seawall, at a rate of up to 20 ha per year. Today, these saltmarshes occupy almost 4,000 ha and are among the widest along Europe's west coast.[15]

Thus, since this bay has existed, it has been physically and biologically transformed for the needs of humans. Since the Bronze Age, the fish of the bay were exploited with fixed fisheries of which most were of a V-shaped timber construction (racks). On the Norman coast, some fisheries built with low stone walls still remain. Shrimps are specifically captured by means of *tésures* (numerous fixed nets laid out together in order to block the way of fish and crustaceans) or through the use of *haveneaux* (bag nets with a shaft), with typical local fishing tools called *dranet* in Brittany and *bichette* by the Normans. The gathering of molluscs was also an important activity in the bay, for example the collecting of cockles and the dredging of oysters through the use of large sailing boats (*bisquines*). These traditional activities have now made way for fisheries which have significantly transformed the landscape of the bay with oyster tables and mussel beds (which have just obtained an AOC).[16] Hunting has always been established in the bay. First practised at dawn or at dusk (*à la passée* or *à la botte*), then also during the night since 1947–1948, with the construction of ponds and huts (*gabions*) on the saltmarshes. In addition, the saltmarshes have been exploited by sheep since the Middle Ages. Grazing modifies the marsh's natural vegetation by privileging 'sheep grass' (*puccinellia*) which forms genuine short lawns. These sheep, which are famous for their gustatory quality, are now known by the name of the area which they transformed—the salt grasslands (*les prés salés*).

Thus for 3,000 years, but particularly since the year 709, the date of the first Christian constructions dedicated to Saint Michel, and all along the rising of '*le phare de la Chré-*

*tienté*, the bay has changed, either because of the exploitation of its natural riches or by developments in the littoral zone, mostly related to the reclamation of land from the sea. This evolution was accentuated during the 19th and 20th centuries as a result of the demands of agriculture and of tourism, with the number of visitors continually increasing. The exploited nature of this area did not affect the UNESCO decisions. Indeed, in spite of, or because of, these changes, the bay shelters an exceptional natural heritage:[17]

(a) It has the largest saltmarshes (4,000 ha) and is the richest in floristic species of the French Atlantic coasts.

(b) For migratory birds, it is an area of international importance for wintering, more especially for Anatidae and waders. In the event of a cold snap, the bay is used as a climatic refuge, being able to accommodate exceptional numbers on such occasions (widgeons especially). For them, as for the Brent geese, the saltmarshes play an essential role as a nursery.

(c) The bay is also an important nursery for many fish species: sole, plaice, ray, bass, mullet.

(d) The bay shelters the largest 'deposit' of polychaete worm (*hermelles*) reefs of all the European coasts. These reefs are built by the polychaete annelid *Sabellaria alveolata*, and the individuals form colonies made from agglomerated tubes of sand and shelly remains.

(e) The bay regularly accommodates several species of marine mammals but two species are present all year round: the large dolphin and the seal calf-sailor (the latter registered as an endangered species).

(f) The Rivers Sée, Sélune and Couesnon are classified rivers for migratory fish (eel, salmon, sea trout, shad and lamprey). The River Sée has the best salmon populations and, more generally, has high-quality pisci-cultural stocks, as opposed to the Couesnon where the quality of the water has been degraded.

## Biodiversity—Hidden Values

For the visitor who contemplates the bay from the top of the Wonder, the 'casket' is always beautiful and everything appears fixed. The landscapes of polders, the result of the keen fight to conquer new territories of the sea, the sumptuous surfaces of saltmarshes, the mud flats which extend as far as the eye can see, and even the alignment of the mussel beds (mussel-breeding stakes) which, like 'Buren's columns' in Paris's Palais Royal, look like an architect's desire to magnify the decoration, all participate in a construction of space that can only make the architectural beauty of Mont-Saint-Michel itself more sublime. Yet so many changes have occurred in less than a century. The recent polders whose wet grasslands constituted one of the greatest wintering places for white-fronted geese in France, have lost 99% of their meadow surface in the last 30 years, and have also lost their water fowl. But the most important feature of this bay is the major transformation undergone by the saltmarshes and in particular the progressive loss of their functions. It is important to emphasise this last aspect as it is too often ignored by the public and the users of these quasi-natural areas. The Millennium Ecosystem Assessment (2005) describes the functions and services provided by the ecosystems. This also arose as a result of the international 'Biodiversity: Science and Governance' conference held in Paris (January 2005) which, in the Paris Declaration, insists that 'biodiversity provides goods that have a direct use-value such as food, wood, textiles, pharmaceuticals ... It supports and enhances the ecosystem services upon which human societies often indirectly depend.'[18]

The direct use-value is easiest to identify even if its importance is largely underestimated in our society. Indeed, if the question 'What are the saltmarshes used for?' was asked of the many visitors to the Mount, the answer which one would obtain immediately would be the same as the local farmers who use the marshes to graze their many herds of sheep, the famous '*prés salés*' of Mont-St-Michel bay. The stockbreeders could refine the response by specifying that in some places they also use them to make hay or to graze cattle and even horses. Grazing fundamentally modifies the vegetation cover of the saltmarshes, which it transforms into relatively short vegetation, dominated by graminaes, the *puccinellia*. This latter plant constitutes the principal food source for wintering widgeons, one of the species targeted by hunters. Another wintering Anatidae, the Brent goose, which is, by contrast, a protected species, is also dependent on the marshes with *puccinellia*, though it prefers the zones grazed by bovines which leave slightly longer vegetation compared to sheep. It is understood that farmers, hunters and nature conservators, because they appreciate the value of what these ecosystems can offer, are involved in keeping these marshes transformed by grazing. Unfortunately, overgrazing greatly reduces what the ecosystem can offer because of its effect of extending the area covered by grass that is too short to accommodate geese and widgeon, and even sheep, in their search for food.

On the other hand, the interest in 'natural' saltmarshes is not obvious. Those in Europe are characterised by vegetative cover that depends on the duration of tidal flooding. It is thus possible to distinguish a pioneer zone, a low marsh, a middle, and a high marsh, each zone being characterised by dominant plants. One of the pioneering species is *salicornia* which is used now more and more as pickles in human food. As one moves towards the mainland, one finds the *puccinellia*, characteristic of the low marsh (the juvenile part of the marsh), the extension of which is of interest when supported by 'engineer organisms' (i.e. sheep). The middle saltmarshes are occupied extensively by a very common species widespread from the saltmarshes of Hyères to the bay of Somme, the obiones (*Halimione portulacoïdes*). This middle marsh with *Atriplex portulacoïdes* appears to be generally of little interest to the users of the bay. The high part is occupied by red fescue and sea couch grass (*Elymus athericus*); it corresponds to the 'mature' marsh, the oldest in the process of terrestrialisation. Ten years of ecological research have shown that the middle marshes, so forsaken, are far from being unimportant if one is interested in the ecological services they are able to fulfil.[19] First, the obiones (*Halimione portulacoïdes*) the dominant species of the zone, are very productive; more than 20 tonnes on average of dry organic matter per hectare and per year up to a maximum production of 36 tonnes, without ploughing, fertilisers or pesticides—whereas a cereal crop needs 140–180 kg of nitrogen per hectare for a production of 10–13 tonnes of dry matter.

Most of the organic matter produced by the obiones (*Halimione portulacoïdes*) is decomposed on the spot (at an equal speed to that in wet tropical forests) thanks in particular to the work of a small shellfish 'chopper' (of type *Orchestia*) and to bacteria, some of which cause the mineralisation of organic matter. This results in the production of organic matter as dissolved and fine particles as well as nutrients (nitrogen and phosphorus) which, exported to the marine environment, serve to enrich the mud flats close to the saltmarshes. This enrichment makes it possible to understand the output of these mud flats in terms of benthic micro-algae, the diatoms. Carried inland again by the flood at the time of the rising tides, these diatoms, alive or dead, and the organic micro-detritus (coming notably from the saltmarshes) make it possible partly to explain why Mont-Saint-Michel bay is able to produce an average of 10,000 tonnes of marketed mussels per year (the primary French centre for mussel beds) and 4,000–6,000 tonnes of oysters—without counting the production of invertebrates, some of which are eaten by migratory birds, in particular the waders. If one more closely observes the functioning of these 'natural' saltmarshes, flooded as they are by less than 40% of the tides each year, one realises that they are visited during the immersion (which lasts less than one hour per tide) by fish like the mullet and the one-year-old juvenile sea bass. Many of these fish arrive with an empty gut. The mullet gorge themselves on diatoms which they take to the bottom of the '*criches*' (a local name given to the tidal channels which drain the saltmarshes). The young bass can return with stomachs filled with *Orchestia* which has a very active role in the decomposition of the obiones (*Halimione portulacoïdes*), as mentioned above. The capture of these small shellfish can account for up to 90% of the growth of the bass during their first year of life.

The grazed marshes, by comparison, lose a large part of their ecological function. Indeed they produce less than 5 tonnes of dry matter per hectare per year, which represents an important reduction compared to the middle marsh production. So, by comparison, this represents a decrease of exported organic material and nutriments to coastal marine water. Consequently they shelter only one reduced population of *Orchestia*, which is detrimental to the juvenile sea bass population. These few examples are enough to explain why Eugene Odum, a renowned specialist in ecology, often used to say 'the saltmarshes are the richness of the sea' to convince developers to stop transforming the saltmarshes of the east coast of the USA into arable land through draining. They also make it possible to understand that while farmers, hunters and all those who perceive directly, visually, can see the 'services' provided by the grazed saltmarshes, the fishermen and the shell-fish breeders cannot realise, through lack of knowledge, that the success of their production, and their incomes, can partly be dependent on these ordinary marshes which thus far have been perceived as an embarrassment rather than an asset. The co-existence of areas of easily recognised usage value and areas providing ecological services may appear simple now that these areas have been identified by researchers.

The situation is quite different with the new threat to the functioning of the salt-marshes. For about 15 years they have been invaded by sea couch grass, a characteristic species of the high marshes which now extends to the low marshes, progressing at a rate of 3.9 ha per year. This invasion calls into question both the functioning of the grazed zones (the plant is not eaten by the sheep or the phytophagous birds[20]) and the functioning of the natural zones—despite high production, the grass produces organic matter that is 'resistant' to decomposition because of its high lignin content,[21] which results in little exportable matter but, on the contrary, in the storage of organic matter and the deposition of sediment, leading to a levelling of the ground supporting the perennial installation of this graminae. It is also more unfavourable to the nursery function for the young bass.[22] Grazing by sheep after crushing the vegetation seems able to contain this invasion locally[23] but the high cost of such operations, experimental at the moment on a game reserve, makes the general introduction of this system of control impossible for an area of more than 1,000 ha. To stop this type of invasion, it is necessary to attack the principal cause, known to be due to the increase in the nitrate and nitrate compound content in the water of the rivers (the Couesnon in particular) and in the groundwater. The nitrate compounds are used by grass to synthesise osmoticum, which allows it to resist salt stress and to progress towards zones more and more frequently flooded by tide.[24] Indicative of a true eutrophication, due to the modernisation of agriculture, the development of the single corn crop and industrial breeding, the invasion by grass can no longer be controlled except in the restricted circle surrounding Mont-Saint-Michel and its bay, but it must bring to mind a new vision of what could become the agricultural production in the bay area with the restoration of the degraded catchment area—the disappearance of the *bocage* and wetlands. In contrast to what might be believed, such an ambition is not a utopia. The major development works undertaken since the beginning of 2006 to give, once again, a maritime character to Mont-Saint-Michel have helped the dialogue between Breton and Norman people and accelerated various initiatives, including those from the Civil Service and from elected representatives, who consider that the future of the bay depends mainly on relevant thinking about Catchment Area Management. These different viewpoints, this co-involvement of the Civil Service, NGOs and ecosystem users, basing their opinions on the long-term research required for the evolution of the ecosystem, appear to be the first step towards what might be called 'the new governance of the ecosystems for the conservation of their biodiversity, their functions and services'.

The case of the Banc d'Arguin National Park, in Mauritania, perfectly illustrates the interest in such thought processes. This park, created in 1976 at the request of such scientists as Theodore Monod, and the French Foreign Ministry, fascinated as they were by the ornithological richness of the place, gained World Heritage List status in 1989. It covers an area of 12,000 ha, of which 6,500 correspond to coastal shallow waters. This area was used as a background for the famous *Radeau de la Méduse* by Géricault. This marine space is the Imraguen's territory, yellow-mullet shore-fishermen, in partnership with 'driver' dolphins. Over-fishing for mullet outside the park and the demand for shark fins led to a difficult situation in the park; the pressure of fishing exerted on the selachians not being easily conceivable in a protected area. The International Foundation of the Banc d'Arguin established lasting relations at the higher level of the Mauritanian state while showing that the Banc d'Arguin was not only a wintering zone for European birds but also the largest nursery on Africa's west coast, and one essential to the survival of fishing for all the nations of this west coast (a priceless service rendered by this ecosystem's complexity). The Foundation also gained the Imraguen's confidence by giving them the means (thanks to Breton carpenters) of restoring their boats, the *lanches* (sailing boats given by Canarian fishermen), and for building some new ones within a co-operative framework encompassing all the fishing villages. The conditions for a permanent debate about the management of the natural wealth of the park were created, which resulted in the recognition, by law, of the exceptional status of this park which depends directly on the General Secretary of the Mauritanian government, on the Imraguen becoming park 'wardens' and taking part in the fight against poaching, and by the organisation of an annual forum joining together the Civil Service, elected representatives, village chiefs and the inhabitants of the park, all of whom are concerned with the management of resources. New rules of management were defined based on the publication of the research results. Thus, in the area of fishing activity, after having set up a one-month biological rest period followed by temporary restrictions on the use of nets for ray and shark, the Imraguen people proposed the closure of shark fishing in 2004, followed in 2005 by the destruction by burning (after receiving grants) of the welded nylon mesh net. The Imraguen people are currently focusing instead on angling, which is much more selective and guarantees quality products.

The results of this type of governorship, though certainly time consuming, are sufficiently convincing to be used as an example of the setting up of new methods of management for protected natural spaces, in particular those that form part of the World Heritage. It is time to think that nature, which across the world shows some evident signs 'of breathlessness', also deserves some major development if future generations are to be able to benefit from the free services that it so liberally dispenses to humanity. But we must not delude ourselves. Nothing can be done without long-term research, taking into account the permanent evaluation of the changes affecting the ecosystems and their consequences, along with analysis of the services provided and of the consequences of management measures taken to preserve natural resources. This is the price we must pay to prepare a better future for later generations, those in the developing countries included.

Translation: Annie Chapon
Comments and corrections: Micheàl O'Curraoin (Ireland).

## Notes

[1] De Lumley, *L'Homme premier*, 11.
[2] Kislev and Hartmann, 'Early Domesticated Fig in the Jordan Valley', 1372–1374; Cauvin, *Naissances des divinité*, 77–88; Vigne, *Les Origines de la culture*, 12–15.
[3] Mazurié de Keroualin, *Genèse et diffusion de l'agriculture en Europe*, 160.
[4] Desroches-Noblecourt, *Lorsque la nature parlait aux Egyptiens*, 44–45.
[5] Ibid, 10.
[6] De Ravignan et al., *L'Atlas de la France verte*, 36.
[7] D. Clavreul, 'Contribution à l'étude des interrelations paysages/peuplements faunistiques en région de grande culture', thèse de doctorat de 3$^{\text{ème}}$ cycle, Université Rennes 1, 1984, 283.

[8]   Lefeuvre et al., 'Rapport de synthèse du chapitre IV', 315.
[9]   D. Clavreul, 'Contribution à l'étude des interrelations paysages/peuplements faunistiques en région de grande culture', thèse de doctorat de 3ème cycle, Université Rennes 1, 1984, 283.
[10]  Binder, 'Ouvrages hydrauliques et entretien des cours d'eau', 106.
[11]  Couvet et al., 'Les Indicateurs de biodiversité', 45.
[12]  Fabre, *Le Pont du Gard*, 71.
[13]  This qualifier actually indicates the Gothic part of the abbey, finished in the 14th century, which is only one element of a unit which was built since the end of the 9th century. Between fires and rebuilding, between wars and times of peace, by successive stacking of buildings of which the oldest preserved is the admirable church of Notre-Dane Sous Terre dating from the 10th century, sumptuous buildings were to be established to crown the rock. Mount-Saint-Michel, however, is also a village and a fortress whose ramparts rest directly on the marine sediments, the *tangue*.
[14]  It is to be noted that if this situation is related to human activities, a retrospective analysis of the bay sedimentation shows that with each negative fluctuation of marine water level, the salt-marshes were 'naturally' replaced by peat bogs and freshwater marsh plants, with any organic matter produced being hidden under the marine sediment during the resumption of the transgression. It is this 'laminated' system which, as a result of carbon dating of the peat, facilitated the provision of a chronology of all the events which characterise the history of this bay.
[15]  Lefeuvre et al., 'European Salt Marshes Diversity and Functioning', I–X.
[16]  Appellation d'origine contrôlée.
[17]  Lefeuvre, 'La Baie du Mont saint Michel et ses bassins versants', 1–47.
[18]  The Paris Declaration on Biodiversity (pp. 304–6) is contained in the acts of the Conférence Internationale 'Biodiversité Science et Gouvernance', 2005, edited by the Muséum National d'Histoire Naturelle for members of the Institut Français de la Biodiversité. Volume 316 pp. + CD.
[19]  Lefeuvre et al., 'European Salt Marshes Diversity and Functioning', 147–161.
[20]  Schricke, *Les Aménagements de la réserve maritime de chasse maritime de la baie du Mt St Michel*, 60–64.
[21]  L. Valéry, 'Approche systémique de l'impact d'une espèce invasive: le cas d'une espèce indigène dans un milieu en voie d'eutrophisation'. Thèse Muséum National d'histoire Naturelle, 2006, 276.
[22]  Lafaille et al, 'Does the Invasive Plant *Elymus athericus* Modify Fish Diet in Tidal Salt Marshes?', 739–746.
[23]  Schricke, *Les Aménagements de la réserve maritime de chasse maritime de la baie du Mt St Michel*, 60–64.
[24]  Leport et al., 'Biochemical Traits Related to the Adaptation to Salinity of *Elytrigia pycnantha*', 31–38.

## References

Binder, W. 'Ouvrages hydrauliques et entretien des cours d'eau: l'expérience bavaroise.' In *Milieux naturels: illustrations de quelques réussites*, edited by C. Henry and J. C. Toutain. Paris: CNRS, 1986.

Cauvin, J. *Naissances des divinité: naissance de l'agriculture*. Paris: CNRS, 1997.

Couvet, D., F. Jiguet, R. Julliard and H. Levrel. 'Les Indicateurs de biodiversité'. In *Biodiversité et changements globaux*, edited by R. Barbault and B. Chevassus-au-Louis. Adpf. Paris: MAE, 2004.

De Lumley, H. *L'Homme premier: préhistoire, évolution, culture*. Paris: Odile Jacob, 1991.

De Ravignan, F., P. Roux, P. Brunet, C. Gay and A. Brun. *L'Atlas de la France verte*. Paris: Jean-Pierre de Monza, 1990.

Desroches-Noblecourt, C. *Lorsque la nature parlait aux Egyptiens.* Paris: Philippe Rey, 2003.

Fabre, G. *Le Pont du Gard: l'aqueduc antique de Nîmes.* Barbetane: Equinoxe, 2001.

Kislev, M. E. and A. Hartmann. 'Early Domesticated Fig in the Jordan Valley'. *Science* 312 (2006): 1372–74.

Lafaille, P., J. Pétillon, E. Parlier, L. Valéry, F. Ysnel, A. Radureau, E. Feunteun and J. C. Lefeuvre. 'Does the Invasive Plant *Elymus athericus* Modify Fish Diet in Tidal Salt Marshes?' *Estuarine, Coastal and Shelf Science* 65 (2005): 739–46.

Lefeuvre, J. C. 'La Baie du Mont saint Michel et ses bassins versants: un modèle d'anthroposystèmes'. In *130ᵉ congrès. Dol 2003.* Vol. CXII. Bannalec: Association Bretonne et union Régionaliste Bretonne, 2004.

Lefeuvre, J. C., V. Bouchard, E. Feunteun, S. Grare, P. Lafaille and A. Radureau. 'European Salt Marshes Diversity and Functioning: The Case Study of the Mont Saint-Michel Bay, France.' *Wetlands Ecology and Management* 8 (2000): 147–61.

Lefeuvre, J. C., J. Missonnier and Y. Robert. 'Rapport de synthèse du chapitre IV: caractérisation zoologique' [Ecologie animale]. In *Les Bocages: histoire, écologie, économie. C.R. Table Ronde CNRS: aspects physiques, biologiques et humains des écosystèmes bocagers des régions tempérées humides.* Rennes, 5–7 July. Orne: Edition E. D. I. F. A. T.–O. P. I. D. A., Echauffour, 1976.

Leport, L., J. Baudry, A. Radureau and A. Bouchereau. 'Biochemical Traits Related to the Adaptation to Salinity of *Elytrigia pycnantha*, an Invasive Plant of the Mont-Saint-Michel Bay'. *Cahiers de Biologie Marine* 47 (2006): 31–38.

Mazurié de Keroualin, K. *Genèse et diffusion de l'agriculture en Europe.* Paris: Errance, 2003.

Schricke, V. *Les Aménagements de la réserve maritime de chasse maritime de la baie du Mt St Michel: bilan du suivi ornithologique et botanique.* Rapport scient. Off. Nat. De la Chasse et de la Faune Sauv., 2004.

Vigne, J. D. *Les Origines de la culture: les débuts de l'élevage.* Paris: Le Pommier, 2004.

# The Evolution of Approaches to Conserving the World's Natural Heritage: The Experiences of WWF

Chris Hails

This paper is very much a personal and institutional perspective on approaches to conserving the world's natural heritage. It reviews the changing perceptions of conservation of the natural world and how one global organisation has reacted to these. It is not intended to be a comprehensive history of the conservation movement.

## Background

The World Wildlife Fund was founded in 1961 (11 September 1961), by a small group of ardent, mostly British naturalists and conservationists such as Peter Scott, Max Nicholson, Guy Mountfort and Julian Huxley. The latter had published a series of articles in the UK *Observer* newspaper on his observations of an environmental crisis in Africa. He received a reaction from the businessman Victor Stolan in December 1960 who proposed the establishment of an international organisation to raise funds for the conservation of wild species. Huxley, Nicholson and companions reacted to this by forming WWF, known then as the World Wildlife Fund, a little under a year later. All of those founders had connections with other conservation organisations such as the International Union for the Conservation of Nature (IUCN), the Fauna Preservation Society, UNESCO, the British Nature Conservancy, etc., and so WWF had a springboard from their knowledge and connections.[1]

Until that time conservation had been largely the domain of scientists and hunters, but WWF moved the agenda out into the public arena for the first time, using publicity and public appeals skilfully.[2] In a post-empire world this primarily emotional appeal to 'help save wildlife' struck a chord with the public and WWF was able to raise significant funds and donated $1.9 million to projects in Africa, Europe, India and other places in its first three years—a considerable sum in the early 1960s.

What is remarkable is the speed with which WWF was able to become established and grow. This was partly due to the well-connected and influential individuals who were associated with the founders. But it may also have been that the 'time was right' for such an organisation. Television was beginning to bring world affairs into people's homes; the post-war industrial boom had raised sensitivities to matters of pollution and waste disposal; and several years of controversy culminated in Rachel Carson's famous book *Silent Spring* which cautioned on the effects of pesticide abuse.[3] The famous 1960s were also a time of pressure on the 'establishment', of non-acceptance of the status quo or traditional solutions to problems. Thus a new approach to a now visible wildlife crisis had its attractions to a wide audience.

This was also a time when the deeper relationships between humans and nature began to be examined. The dependence of our ancestors on wildlife stocks to hunt and fish had always been recognised; those days were long gone, but a popular late 1950s feeling that the resources of the sea were limitless was being replaced by mounting suspicion that things were not that simple. Evidence of loss of topsoil, water shortages and pest outbreaks created by industrial-scale farming was giving rise to broader questioning of human relationships with the environment, and it was during this period that James Lovelock's Gaia Hypothesis was formulated.[4]

Thus WWF was founded for specific purposes during a period of wide-ranging thought. It was established as a Swiss Foundation registered in Zurich, and the deed of foundation specified amongst the purposes of the organisation '… the conservation of world fauna, flora, forests, landscape, water, soils and other natural resources …' This far-reaching vision for WWF is perhaps even more relevant today than it may have been in 1961, because people's attention was drawn by an emotional argument based upon the preservation of charismatic species. Reflecting this, the British appeal of WWF was launched with pictures of black rhinos in Africa under the headline 'doomed', and Peter Scott had taken George Waterson's sketches of the giant panda Chi-Chi, then residing in London Zoo, and turned it into the logo of the organisation. Chi-chi was the only giant panda residing in the West, had arrived from the mysteries of communist China, and was an evocative species symbol for the challenges facing those concerned with the preservation of wild nature.

So despite some deeper thinking which underpinned it, the early days of WWF were ones which were dominated by a preservationist agenda for species and habitats, based on popular appeal.

## The Next Generation

This approach ran successfully through the 1970s while, along with the spread of television (soon to be in colour) and the growth of wildlife documentary films, public awareness of conservation and natural heritage issues grew exponentially. But with that awareness came the realisation that a rather crisis-driven, spotty approach to conservation was not achieving the long-term solutions that were sought, and that economic development continued to impact heavily upon nature. In 1980, WWF came together with IUCN and the newly formed UNEP to produce the modestly named 'World Conservation Strategy'; at the time this was a landmark document because it linked human activity, human well-being and its dependence upon nature all as one. It stressed the interrelationships between conservation and development and first gave currency to the term 'sustainable development'.[5] Conservation had suddenly become much more complicated but much more relevant to the modern world.

The 1980s were marked by a closer examination of development issues and their relationships to the environment. In 1985 WWF formally re-registered its name as World Wide Fund for Nature, to try and escape the preservation of animal species image and reflect a broader view of the situation. In 1987 the World Commission on Environment and Development (WCED) produced *Our Common Future* and this properly defined 'sustainable development'.[6] Most significantly the UN began the planning for a World Conference on Environment and Development ('the Rio Summit') for 1992. In advance of Rio, and now a decade further on, IUCN/UNEP and WWF once again came together to produce *Caring for the Earth: A Strategy for Sustainable Living*[7] which explored from a strategic perspective how the concept of sustainable development could be implemented in practice.

All this activity served to move environment and conservation on to a higher plane. It was no longer the specialised interest of scientists, hunters and animal lovers; there was a realisation that a sound environment was the starting point for all human development and welfare and that our activities were inextricably woven into the milieu in which we live.

## The WWF *Living Planet Report*

As espoused in *Caring for the Earth*, WWF began to take a much more strategic approach to its conservation activities, and also wanted to explore the linkages between nature and human activity by looking at the state of nature and how it was changing. Businesses and economies had their own barometers of change in the form of the Dow

Jones, CAC40 and FTSE indices. These could be used to see how the world of commerce was changing; would it be possible to do the same for nature?

In 1998 the first WWF *Living Planet Report* (LPR) was published,[8] containing estimates of the changing state of nature based on changing populations of vertebrate animals; it also contained an estimate of human pressure on the planet. Eight years later the LPR is now a bi-annual publication; the LPR 2006 has grown to contain the Living Planet Index, a composite of data from 3,600 species populations, and also the ecological footprint—an index of the area of the planet needed to sustain human activity (see Figure 1). These two indices show that, roughly over the past 30 years, the natural world has lost approximately 30% of its health as indicated by declining populations of wild species, whilst at the same time human activity has caused our ecological footprint to more than double during the same period. The causal relationship between the two is not difficult to deduce.

In reviewing the ecological footprint we can learn even more, because it shows that some time in the mid-1980s human activity passed the point that the planet can sustain and that we now exceed it by about 25%, in other words the human population requires 1.25 planets to sustain present levels of consumption. At the moment the human population of the planet can live on more than is available, because we continue to 'mine' the accumulated capital of such things as stocks of timber from forests, or fish from the sea, and we also take the products of past millennia from the ground in the form of fossil fuels. This can be likened to spending more money than one earns each month by draining savings from the bank account. However, this cannot continue and humankind continues to degrade the planet.

## The Challenge

Given a new understanding of the relationships between economic development and its impact and its dependency upon the environment, what should be the best approach for a private conservation organisation such as WWF? Although of significant size (WWF will spend nearly $500m on conservation in the current fiscal year), the total amount of activity even with this large sum is seemingly trivial when compared with the total economic activity of the world—global GDP in 2005 was of the order of $60 trillion. Thus the challenge how to influence the rate of degradation of our natural heritage and to influence development paths on to a more sustainable trajectory.

The work to be done falls into three main areas:

(1) Direct biodiversity conservation: like the good mechanic dismantling an engine, this involves 'keeping all the pieces'—ensuring that species are not lost or ecosystems irretrievably damaged by the threats.
(2) Reversing the threats: this involves tackling the immediate cause of environmental decline—those threats which are direct such as over-fishing, deforestation, or illegal wildlife trade, and those which may be more indirect such as climate change or toxic pollution.

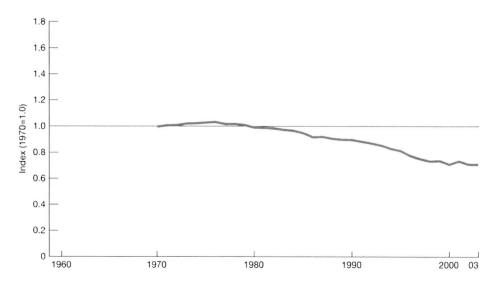

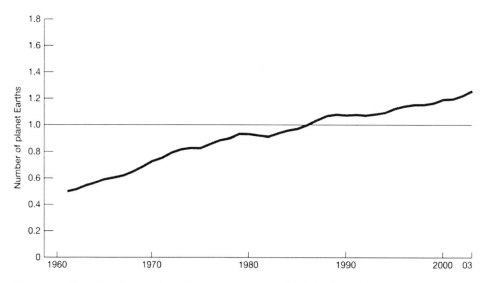

**Figure 1**    Top: the Living Planet Index 1970–2003 which declined by approximately 30% during this period. Bottom: the ecological footprint related to the biocapacity of the planet. Some time in the 1980s the ecological footprint passed the biocapacity of the planet and is now in overshoot. (*Source*: WWF, *Living Planet Report 2006*.)

(3) Creating favourable or 'enabling' conditions: many of the threats to natural heritage security exist because of fundamental failures in the policy frameworks and decision-making processes that influence economic trends and development paths. Influencing those so that they encourage environmentally sound behaviour is essential to cure the underlying disease causing environmental decay.

**Rising to the Challenge**

During this period of more sophisticated understanding of environmental challenges, over the past 20 years there has also been significant growth in non-governmental organisations (NGOs) addressing conservation issues, especially at an international scale. In some instances this was reflected in the growth of existing institutions; WWF grew from about 25 major offices in the mid-1980s to nearly 60 by the end of the 1990s and with activities in more than 100 countries. Long-established NGOs such as the US-based The Nature Conservancy (TNC) began to migrate from traditional land-owning US-based conservation to increasing engagement in developing countries; a move also reflected by the Wildlife Conservation Society—a conservation research and extension activity of the Bronx Zoo. In other instances whole new organisations developed such as Conservation International which started in the late 1980s with an international mandate from its inception.

In all of these organisations there was a growing frustration that we were winning many small battles but still losing the war; on what scale we were losing was only properly realised when WWF produced its first Living Planet Index in 1998. But each in their own way was struggling to produce bigger and better results with the resources at their disposal.

Some institutions focused on particular themes—thus the World Resources Institute (WRI) 'Forest Frontiers' programme began; others focused upon the danger of extinction of species—Conservation International expanded upon the Norman Myers 'hotspots' approach. In WWF and TNC this meant working on a larger geographic scale and the concept of Ecoregion Conservation was developed. For WWF this was based upon an analysis of the global distribution of biodiversity resulting in a map now known as the 'Global 200'.

**The Global 200**

The Global 200 began with WWF asking the question: 'if we wish to conserve biodiversity, where should we be investing our precious conservation funds?' It was an exercise in prioritisation, recognising that if the available resources were spread too thinly they could not achieve the desired result; then how should they be focused?

WWF scientists gathered published data on the way in which species were distributed across the planet. Those distributions coalesced into patterns they called 'ecoregions'—an ecoregion is defined as a large area of land or water that contains a geographically distinct assemblage of natural communities that (a) share a large majority of their species and ecological dynamics, (b) share similar environmental conditions, and (c) interact ecologically in ways that are critical for their long-term persistence.[9] These ecoregions had reasonably well-defined boundaries and could be plotted on a map. To turn this into data which could help determine conservation priorities, WWF selected the approximately 200 ecoregions (actually 238) which best represented the distribution of biodiversity on a global scale, and so resulted the 'Global 200' (see Figure 2). This analysis recognised for the first time that it was not

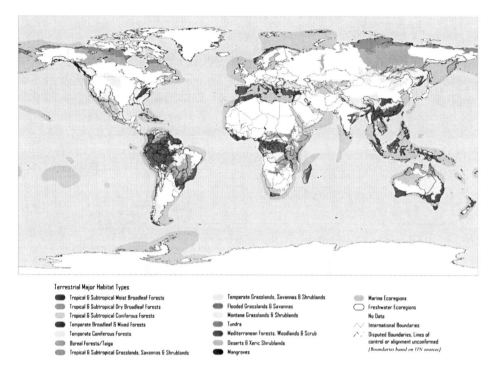

Terrestrial Major Habitat Types

Tropical & Subtropical Moist Broadleaf Forests
Tropical & Subtropical Dry Broadleaf Forests
Tropical & Subtropical Coniferous Forests
Temperate Broadleaf & Mixed Forests
Temperate Coniferous Forests
Boreal Forests/Taiga
Tropical & Subtropical Grasslands, Savannas & Shrublands

Temperate Grasslands, Savannas & Shrublands
Flooded Grasslands & Savannas
Montane Grasslands & Shrublands
Tundra
Mediterranean Forests, Woodlands & Scrub
Deserts & Xeric Shrublands
Mangroves

Marine Ecoregions
Freshwater Ecoregions
No Data
International Boundaries
Disputed Boundaries, Lines of control or alignment unconfirmed
*[Boundaries based on UN sources]*

**Figure 2**    The WWF Global 200 ecoregions; terrestrial ecoregions are shaded according to the Major Habitat Type. (*Source*: Olson and Dinerstein, 'The Global 200'.)

only coral reefs and rain forests that were important, but that deserts, Mediterranean regions, and the tundra contained unique species which, if lost, could never be replaced.[10]

This mapping approach clearly indicated where the work should begin. But it created a new problem: each ecoregion demanded working at a scale which conservationists had never tried to work at before, but which in fact was probably in better relation to the threats they were facing. So a new challenge now resulted—how to work at an ecoregional scale.

## Ecoregion Conservation

At the same time that WWF was identifying the Global 200, a sister organisation—TNC—was also examining how to work at a large geographical scale. Between the two they invented a new approach which they called 'Ecoregion-based Conservation'. This was quite an exciting period as conservationists had not developed a new tool for their problems for some years, and here was a new and ambitious approach in the run-up to the new millennium.

Ecoregion conservation basically involves standing back, as if from space, and asking 'what needs to change to secure the long-term conservation of this ecoregion?' The approach is based upon four fundamental principles of biodiversity conservation:

(1) representation of all native habitats;
(2) maintenance of viable populations of all native species;
(3) maintenance of essential ecological processes;
(4) maintaining resilience to ecological change.

By viewing these needs from a distance the observer is forced into thinking about the fundamental changes that are required to achieve them and the challenge creates questions that begin to give the clues as to the work which is required:

(a) What are the current trends of environmental change within the ecoregion and who is affected by them, both positively and negatively?
(b) What current processes are taking place within the ecoregion, especially related to development, and how might they be impinging upon the environment? This usually entails a host of issues such as landscape change for agriculture, industrial development, city expansion, port construction, change in drainage patterns through dams, irrigation channels or other hydrological change. What can be done to mitigate the impact of these?
(c) What are the fundamental forces driving those changes which may be damaging? This may be economic pressures from inside the ecoregion or outside the ecoregion (e.g. structural adjustment loans, or perverse subsidies driving change), demographic issues, internal political issues.
(d) Who are the players concerned with the environment and what are their capacities to deal with the challenges? This involves looking at both government and non-government institutions and their strengths and weaknesses.
(e) What are the key landscapes and habitats in the ecoregion and do they have adequate protection currently? If not, what should be added to a protected areas system?

By standing back and answering far-reaching questions like these, ecoregion conservation forces the questioner to think broadly and creatively, to look at the wider picture, to examine fundamental drivers rather than immediate symptoms. It requires that not only the systems for protecting nature (national parks and protected areas) be adequately addressed, but that policies influencing them and the land area which connects them are also sound. These policies may be those of governments inside the ecoregion (e.g. land-use policies, water-quality policies, transport plans, inter-ministerial relationships, etc.), or they may be policies stemming from institutions outside the ecoregion—the impact of World Bank structural adjustment loans, EU agricultural subsidies and how they influence agriculture in developing countries, foreign direct investment and how it impacts poverty-alleviation programmes and trade in various commodities. All these may be of fundamental importance to environmental security within an ecoregion, but may require work in centres well away from the specific ecoregion.

By asking who is involved, who are the environmental stakeholders, an ecoregion approach also encourages the formation of partnerships to work together on a conservation programme. This latter point is critical as normally a single organisation on its own cannot cover the whole range of activities which are needed; it is usually essential to reach out to others with different skills, interests and needs. Some of the most important conservation breakthroughs of recent years have resulted from the joint activities of non-traditional partners, both of which may have had an interest in a sound environment but perhaps for different reasons. This issue is addressed below under the heading 'The Changing Role of Commerce'.

## Thinking Globally

The Global 200 and Ecoregion Conservation enabled WWF to focus its attention on some of the most globally significant parts of the planet and to address environmental change in an holistic manner. However, a purely geographical approach would have missed some of the important global processes underway during more than a decade of globalising economies and the weakening of international boundaries. Starting in the early 1990s an increasing permeability of international borders resulted from a variety of factors including increasing liberalisation of trade, high-speed communications in the Internet age, a burgeoning of (especially multinational) corporate power and a weakening of government authority, in a bundle of symptoms loosely described as 'globalisation'.

Whilst this process of globalisation stimulated trade and commerce, and brought increasing wealth to millions, not all of this activity was of benefit to the environment. Increasing commercial activity brought growth in resource consumption, not only to provide raw materials but also to meet the demands of the beneficiaries who now had greater buying power. It also brought a widening gap between those caught up in commercial prosperity and those not so engaged. This widening gap between rich and poor culminated at the 2002 UN World Summit on Sustainable Development in Johannesburg in the establishment of the Millennium Development Goals, which set an agenda for lifting people out of poverty.

Clearly the marketplace had not only done insufficient for the poor and disenfranchised, but as we can now see from the LPR it has also failed the environment. The 2006 LPR contains a graph which shows the relationship between the ecological footprint and the UN Human Development Index (see Figure 3). This shows that the development trajectory of most countries bypasses the criteria for sustainability. This then presents yet another challenge for those concerned with the conservation of the world's natural heritage: how do we turn the juggernaut of the world economy into a direction that favours the environment? Globalisation and world trade is not something that one can be 'for' or 'against', it is a fact of life, an inevitable force which we need to direct towards sustainability. WWF recognised this some years ago and has been establishing various mechanisms which could lead to a marketplace move to sustainable behaviour. The most successful to date has been relating to the timber trade.

Forests worldwide are in decline as a result of the over-harvesting of timber. The wealth-driven growth of the construction and furniture industry and the growing demand for pulp and paper have put enormous strains upon supplies from forests. In

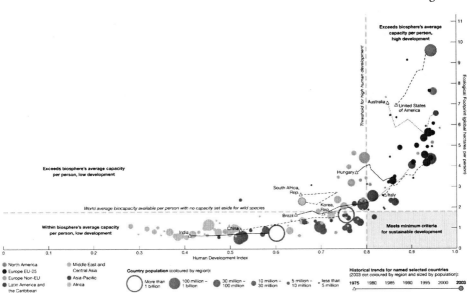

**Figure 3** The ecological footprint and Human Development Index. An ecological footprint of 1.8 global hectares or less is within the per capita biocapacity of the planet; a Human Development Index of 0.8 or greater is considered to be 'high development' (UNDP). Most countries with high development have already exceeded the per capita biocapacity and cannot be considered as sustainable. (*Source*: WWF, *Living Planet Report 2006.*)

the temperate zones of Europe this has been recognised and some modest increase in forest areas has resulted from the establishment of large plantation schemes (but only after most of Europe's forests had already been destroyed). However, in some temperate forests (e.g. the Pacific coasts of Canada and the USA), and broadly in tropical areas, the battles between various forest interests have sometimes been fierce. In the early 1990s the concept of third-party certification for sustainable timber production was established under the name of the Forest Stewardship Council (FSC). The FSC was established as an accreditation agency which could verify country-specific certification systems following the FSC standards and criteria for environmentally and socially sustainable forest management systems. A piece of timber carrying the FSC logo could carry with it the assurance of sustainability—a 'light footprint' in the language of the LPR.

However, for a market mechanism to be effective there has to be demand as well as supply. Thus, by creating a momentum through public and consumer education and awareness programmes, WWF created a new demand for wood with the FSC logo, and groups of timber traders became committed to trading in sustainably produced timber. These timber companies came to realise that continuing environmental decline would inevitably lead to stricter regulations, public demand for action, and difficulties with supplies. Their change in behaviour was not entirely altruistic (although the concept of

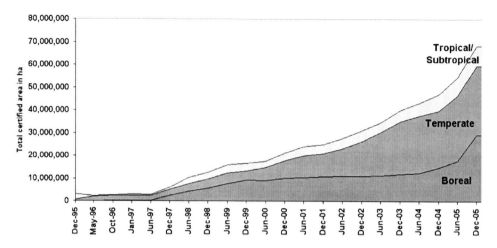

**Figure 4**    Growth in the area of production forests certified under the Forest Stewardship Council (FSC) scheme, 1995–2005. (*Source: WWF.*)

corporate social responsibility (CSR) has now emerged) but also made good business sense for them as well. Approaching the end of 2006 there are over 70 million ha of production forests certified under the FSC scheme with activities in 72 countries (see Figure 4).

The growth of FSC is interesting from a natural heritage conservation point of view: it is a long way removed from the traditional approaches to conservation, it is deeply rooted in international commerce, yet its success could have far-reaching consequences for forest integrity and biodiversity conservation. What is more important, although WWF was a major player in the development and launching of the idea in the first instance, it has now become a self-sustaining force related to the timber industry— a new way of doing business which no longer requires the strong intervention of an NGO.

For a conservation organisation this is important. Referring back to earlier comments on the impossibility of tackling all the environmental needs of the world, it is vital that conservation NGOs find ways of instigating sound practices that can then become self-sustaining, so that the organisation can move their limited resources to a new challenge.

### The Changing Role of Commerce

Moving forward with these ideas WWF has then applied them to the sea. The UN Food and Agriculture Organisation told us many years ago that 75% of the world's fisheries were either depleted or over-fished. Fishing effort was continuing to rise whilst catches were stable or declining. Not only this but species were crashing as populations passed the threshold of sustainable off-take—cod in the North Atlantic, for example, whilst new and strange species were appearing in our shops such as hoki and pollack, as previous commercial species became rarer and more expensive for the average consumer. Into this arena WWF introduced the Marine Stewardship Council (MSC)—the fish equivalent of the FSC.

Interestingly, when WWF launched the MSC it did not do so alone, but with strong support from and close co-operation with a major multinational company (Unilever) which at that time was one of the world's largest consumers of fish. In the 1970s and 1980s such a company would have been perceived as the enemy of the conservationists, but was now a vital part of the solution.

This is a good demonstration of the importance of partnerships between stakeholders who may have quite different reasons for achieving a common result; it is equally applicable on a global scale as it is on an ecoregional scale. If a company or institution has a business interest in a particular resource for the processes it is involved in, then it can be moved from being an exploiter depleting that resource to a user protecting its supplies. This change can be seen in many areas beyond timber and fish; agricultural operations that recognise natural enclaves help keep pests off their land; or drinks companies that need to defend clean water supplies; also as environmental perturbations increase we are seeing insurance companies and financial institutions increasingly concerned with the risks associated with global climate change—probably the biggest single challenge the conservation community faces.

In these ways there now opens up a multiplicity of new avenues of co-operation to the benefit of the world's natural heritage, with sometimes rather unusual bed-fellows working together for a common result.

## A Strategy for Natural Heritage Conservation

A sound institutional strategy is essential to mobilising a global network such as WWF, in the face of apparently insurmountable challenges. In WWF a strategy was built around the concept of the Global 200 ecoregions providing a geographical focus and on-the-ground experience of what it takes to deliver environmental solutions, along with a set of global issues such as the FSC and MSC described above. These came together as a set of priorities for the organisation for which specific targets were established, and measurement systems put in place to monitor progress towards those targets. These became the guiding lights for all the branches of the organisation, enabling each branch to engage in a global effort that could create change for the better. Each set of activities was rooted in a national context so that the solutions could be delivered in a locally appropriate manner, which greatly increased their uptake and probability of success. This helped to build teamwork on a global scale.

However, this should not be regarded as a 'blueprint' or perfect solution for the world's environmental challenges. Conservation is too much a high-risk business to assume that there is one perfect answer. Just as the world is dynamic so conservation organisations too have to be dynamic, constantly looking for new opportunities and new solutions, reacting to change in the way the world operates, engaging that change and adapting strategies accordingly.

## Conservation Organisations as a Facet of Civil Society

The inception, development and impact of organisations such as WWF are a reflection of the relationship between people and nature. Its very existence is a symptom of concern amongst certain sectors of society, driven by the apparent lack of concern by others.

Human societies contain an almost endless variety of values, and for many of these values to involve nature is not surprising. For millennia nature shaped human culture. The forces of nature shaped the evolution of *Homo sapiens* to what we are today. Those same forces also shaped human society. The mastery of centralised agriculture gave human societies the luxury of organising social structures by releasing certain individuals from the need to be food providers. But it was nature that provided the wild species for domestication, and it was nature that maintained soil structure and fertility, that provided water, and until recent years predictable seasonal climates for crops to grow. It was also nature that provided the diseases that damaged domestic produce and killed people. It is thus hardly surprising that many rites, rituals, taboos and traditional belief systems are related to both the positive and negative forces of nature in the hopes that appeasement will create a benign result. Some of these traditions are aimed specifically at prevention of over-exploitation of resources. In this way nature has determined many aspects of our culture.

Ironically, as human understanding of the environment grew, and as we became more skilled at managing certain aspects of nature, then so grew a new belief, in the 20th century, that nature could no longer set the limits for human societies. Culture now began to shape nature. Thus businesses, agriculture, forestry and modern 'hunting' in the form of commercial fishing began to push to the limits our relationships with natural cycles and nature's production. This was aided by artificial nutrients and chemical inputs which improved the response of many species and supported the idea that there were no limits to our exploitation of nature. Thus our culture began to shape nature.

We now know that this was not the end of the story; whilst nature lost the battle with culture for a while, the impact was felt. The LPR quantifies this, and our experiences with extreme weather events, with rivers that no longer reach the sea, with crashing fish stocks, and forests that once never burned now doing so on an annual basis, all cause us to live the environmental consequences of the maverick approach that certain sectors of our society adopted. This being the case, it is probably quite normal that the part of our society that is concerned by these experiences should invent and maintain an environmental ethic and activist movement, of which WWF is one manifestation. Culture once again responded to nature.

The interest of people in nature is also reflected in the structure of WWF as an organisation. Although it has one name and one 'brand' and all the consistency of the corporate world that goes with such things, WWF also has several million members spread around the world, and its very composition is one of semi-autonomous organisations which can build a strong local identity and reach out and tap into the way conservation is best manifested in each country. This is important as, for the reasons already mentioned, relationships with nature involve strong cultural ties, and the way concern is expressed is in the form determined by the culture of the country. This 'people's network' approach has gained strength in recent years through the Internet age. At the click of a button, last year about 4 million people visited WWF's website and conducted 11.5 million 'pages views' of information to learn immediately what is happening in the world. The same medium has enabled more than a million people to take direct action online to lobby decision makers to move in the right direction, and to congratulate those that already had. It is remarkable that this modern technology has enabled individuals in our modern society to have even greater engagement than ever before.

In an iterative process our culture will continue to shape our attitudes to nature, just as nature's response to our attitudes will continue to shape our culture.

## Looking to the Future

As time advances the world will inevitably look more closely at the natural environment, we will begin to talk less about 'conservation of species and habitats' and more about environmental security. Those species and habitats are the very fabric within which our own human lives are woven. For thousands of years anthropologists have shown us how human societies close to nature have recognised their interdependence and have evolved cultural practices and taboos that were kind to nature and ensured the sustainability of their lives, long before the word 'sustainable' was invented.

In our modern, over-engineered society we have lost sight of that interdependence and the fabric is beginning to fray. But out of this and the associated problems the cycle is slowly turning. Climate-induced disasters are making us realise that we are unable to control our environment, but that we must live within its constraints. The year 2005 gave the world the Kyoto protocol—the first small step in the international effort to combat the worst of the threats to our security. The same year also gave us hurricane Katrina which, far from being the most powerful hurricane we have seen, demonstrated most aptly that we cannot engineer the natural world and that if we assume too much then our cities run the risk of being laid waste.

The future is not gloomy, however. The solutions exist, and for the few that don't we are an incredibly creative species. The political will is often lacking, but democratic processes can create that also. The commercial world has learned that it cannot simply take *ad infinitum*. With creativity, understanding and co-operation it will be possible in future to enjoy a secure environment and a high-quality lifestyle both at the same time.

The ultimate goal of WWF is to 'build a future in which humans live in harmony with nature'; a long time ago this was the case. Will we ever get back there again? Perhaps not, but there will always be an active sector of human society which will try to regain the lost ground.

## Notes

[1]  Holdgate, *The Green Web*; Pearce, *Treading Lightly*, 40.
[2]  Adams, *Against Extinction*, 311.
[3]  Carson, *Silent Spring*.
[4]  Lovelock, *Gaia*.
[5]  IUCN/UNEP/WWF, *World Conservation Strategy*.
[6]  WCED, *Our Common Future*, 400.
[7]  IUCN/UNEP/WWF, *Caring for the Earth*, 228.
[8]  WWF, *Living Planet Report 2006*, 40.
[9]  Dinerstein et al., *A Workbook for Conducting Biological Assessments and Developing Biodiversity Visions for Ecoregion-based Conservation*.
[10]  Olson and Dinerstein, 'The Global 200'.

## References

Adams, W. M. *Against Extinction: The Story of Conservation.* London: Earthscan, 2004.
Carson, R. *Silent Spring.* Boston: Houghton Mifflin, 1962.
Dinerstein, E., G. Powell, D. Olson, E. Wikramanayake, R. Abell, C. Loucks, E. Underwood, T. Allnutt, W. Wettengel, T. Ricketts, H. Strand, S. O'Connor and N. Burgess. *A Workbook for Conducting Biological Assessments and Developing Biodiversity Visions for Ecoregion-based Conservation.* Washington, DC: WWF, 2000.
Holdgate, M. *The Green Web: A Union for World Conservation.* London: Earthscan, 1999.
IUCN/UNEP/WWF. *World Conservation Strategy: Living Resource Conservation for Sustainable Development.* Gland: IUCN/UNEP/WWF, 1980.
———. *Caring for the Earth: A Strategy for Sustainable Living.* Gland: IUCN/UNEP/WWF, 1991.
Lovelock, J. E. *Gaia: A New Look at Life on Earth.* Oxford: Oxford University Press, 1979.
Olson, D. M. and E. Dinerstein. 'The Global 200: A Representation Approach to Conserving the Earth's Most Biological Valuable Ecoregions'. *Conservation Biology* 12, no. 3 (1998): 502–15.
Pearce, F. *Treading Lightly: The Origins and Evolution of WWF.* Banson, 2004.
WCED. *Our Common Future.* World Commission on Environment and Development. Oxford: Oxford University Press, 1987.
WWF. *Living Planet Report 2006.* Edited by C. J. Hails. Banson, 2006.

# A Bridge over the Chasm: Finding Ways to Achieve Integrated Natural and Cultural Heritage Conservation

David Harmon

Nature conservation has always been a tough business. Victories are always provisional, battles have to be fought again and again, and progress is often difficult to discern on anything other than a local scale. The difficulties are more acute today than ever. Scientists and environmentalists working to protect nature face a host of well-documented challenges, all of which ultimately stem from the dominion established by people over the Earth and its ecosystems. The litany of looming catastrophes—mass extinctions, out-of-control invasive species, systemic transformations driven by global climate change, rampant air and water pollution, wholesale habitat destruction, and so on—is both well known and, increasingly, well documented. Human activity has mounted to the point where we have literally transformed the planet. Within the last half-century or less, we have arrived at an unprecedented moment in history: there is now no place so remote that it escapes, in the words of Wallace Stegner, 'the marks of human passage'.

Awareness of this fact is widespread among nature conservationists, and has induced a variety of reactions. On a purely emotional level, the response, it would not be too strong to say, is one of anguish. People who care about saving nature tend to care very much; there are few dilettantes among the ranks. Such people feel deeply, even viscerally, distressed at the predicament we have created for the non-human world. For many, the caring is expressed in terms of nature's sacredness; for them, the drive to protect nature is an act of spiritual nourishment and redemption.[1] At the same time, there is an even more complex reaction going on at an intellectual level, one which itself is upsetting to conservationists in a different way. It is the realisation that *nature* and *natural*, as concepts clearly distinguishable from the human activities that come under the umbrella signified by the terms *culture* and *cultural*, are no longer sacrosanct First Principles, immune from critique. It could hardly be otherwise, given how all-pervasive the influence of humans has become, but added to this factual basis, and considerably compounding it, is a theoretical assault on the claimed objectivity and implied supremacy of Western scientific ideals. And, of course, these are the very ideals that are the lynchpin of modern organised nature conservation.

This development is in no small part ironic. Science has been disenchanting the world for well over 300 years, striding from achievement to achievement in fields ranging from astronomy to ecology to microbiology. Now the tables have been turned, and the great disenchanter is itself being disenchanted. Scientists, human beings that they are, have for the most part not liked it one bit. Conservation scientists are no exception.

Some of the contemporary debate can be traced back to the late 1970s and early 1980s, the years of initial reaction to the biologist Edward O. Wilson's extension of sociobiological principles to humans.[2] Social scientists were quick to denounce Wilson's proposed 'macroscopic view' of *Homo sapiens* in which 'the humanities and social sciences shrink to specialized branches of biology; history, biography, and fiction are the research protocols of human ethology; and anthropology and sociology together constitute the sociobiology of a single primate species'.[3] All sorts of objections were lodged. Some argued that a reductionist research model unduly compresses understanding of the full range of human behavioural diversity; others, that there can be no measurable unit of culture transmission analogous to genes; still others, that the deterministic inferences capable of being drawn from human socio-biology rule it ethically out of bounds. An important and recurring criticism held that all this amounts to uninformed poaching on the private property of the social sciences. Thomas Rhys Williams spoke for many in his profession:

In short, anthropologists feel put upon by the growing numbers of evolutionary biologists, population geneticists, assorted types of ethologists, zoologists, and astrophysicists who have turned in recent years to devising schemes and offering generalizations concerning the evolution of human biology, culture, mind and human nature without knowing an *Australopithecus robustus* from *Homo habilis* or Iroquois cousin terminology from a bifurcate collateral kinship system. Put another way, anthropologists tend to be made uneasy by the way persons trained in other disciplines now seem to be wandering selectively through the data of anthropology, while ignoring generally accepted methods, theory, and concepts for the use of such materials.[4]

In the field of conservation most familiar to the author, that of protected areas, this jurisdictional complaint echoes on. It is the clearly detectable, if often unstated, subtext of what one may call the 'social science critique' of protected areas, which has been building over the past 25 years or so. It was given impetus by the publication in 1980 of the World Conservation Strategy,[5] which was the first international prescription to assert that conservation cannot be achieved without joint commitments to alleviate poverty through economic development, and by the third World Parks Congress in 1982, the first to be held outside the USA, whose tenor was strongly towards a more inclusive attitude to humans in protected areas.[6] Paige West and Dan Brockington, both anthropologists, succinctly recapitulate the highlights of the social science critique:

- Too often, natural scientists are indifferent to, or ignorant of, the social context in which conservation takes place.
- Similarly, conservationists fail to see that protected areas are not value-free inscriptions on the landscape, realising self-evident propositions with which everyone should agree, but rather are value-laden prescriptions—politically charged attempts to define how humans should behave in a particular location.
- Even when they accept that cultural considerations need to be accounted for in their plans, conservationists tend to isolate and then try to promote specific cultural practices that support their agenda, rather than seeing these practices as part of a wider social fabric that cannot simply be cut up into a custom-tailored suit.
- By their very success, which hinges on widespread acceptance of their specialness, protected areas tend to devalue more prosaic and mundane ways of experiencing nature.
- Global lists of protected areas do not include non-governmental protected areas, such as privately run game farms or community-conserved areas, that have no official recognition but are important nonetheless.
- Protected areas have unintended, unwanted consequences. These range from delicate social adjustments, as when people claim a new cultural affiliation in order to gain access to resources within protected areas that have been preferentially reserved to a particular group, to the more directly observable, as when they become staging areas for drug trafficking.[7]

The response from the protected-area community to the social science critique has been varied, with some accepting its basic contentions, if not the whole package, and others stoutly resisting. The overall situation is now very much one of contentious flux.

Around the mid-1990s a reaction set in against the trend towards community-based conservation. For example, there were calls for a 'new protection paradigm' that reaffirms the need for more hard-bordered, nationally administered conservation areas actively defended against human intrusions in order to protect biodiversity.[8] These developments have been dubbed 'back to the barricades' or 'fortress conservation' by social scientists and others who promote the social science critique and community-based conservation.[9]

So it would seem that prospects for *rapprochement* are not very promising. Yet despite their understandable reluctance to embrace criticisms that cut to the very foundations of their endeavours, relatively few conservationists today believe that they can be successful simply by applying more and better science to environmental problems.[10] Rather, most recognise that, on some level, conservation science must engage other forms of understanding if progress is to be made. Consider these remarks from the editors of *Conservation Biology* as they introduce a special state-of-the-science issue celebrating the 20th anniversary of this influential journal:

> The most repetitive message coming out of [these retrospective papers] is the great need for interdisciplinarity and inclusion of the various social sciences. This is an obvious imperative; we have to learn how to transform our scientific knowledge into practice. We are facing a fundamental problem relative to human behavior, and the solution ultimately will need to take human behavior into account. This is the great challenge that confronts us in the next decades. Those who still think that biology and ecology alone are sufficient for our task—that good science by itself will save the day—are as much in denial as those who say there is no environmental crisis.[11]

Significantly, there is willingness on the social sciences side to work across the chasm too. For example, West and Brockington do not conclude their critique with a call for a fundamental rethinking of the wisdom of creating protected areas; still less for a wholesale abandonment of the concept. Instead, they ask for basically the same thing as the editors just quoted: better coordination and understanding between conservation biologists and anthropologists. They are careful to point out that when social scientists 'talk about nature and the environment as socially produced', it 'does not mean there is no real, material world out there to be altered, destroyed, restored, or conserved'.[12] From the point of view of natural scientists, this kind of reasoning is both judicious and welcome, in several ways. First, it states that nature is real, as real as anything that that word can signify, and that it is material, made of the kind of ontological stuff that can be studied by science. Second, it affirms that nature is 'out there', distinct (if not always clearly separable) from humans and our interests, actions, desires, values, and so forth. Third, it grants that humans can actually physically destroy this thing called nature, which would not be the case if it were simply a thought construct. Finally, it states that nature restoration and conservation are indeed possible—from which we can reasonably infer that restoration of those elements of nature that have been destroyed or unacceptably altered, and the conservation of what remains intact, are also desirable.

So in the temperate regions of the debate, away from the extremes of either side, there is cause for optimism that a more comprehensive approach to conservation, one which views natural and cultural heritage as complementary rather than conflicted, is achievable. But why even bother to search for commonalities? Is there really any evidence that a truly interdisciplinary approach to the conservation of natural and cultural heritage, one that takes both into account consistently and systematically, will actually produce synergistic, and therefore presumably more effective, results?

This question remains undecided, but there are two lines of reasoning that begin to lead towards an answer. One is primarily inferential, stemming from the general observation that things are getting worse. If one could undertake a comprehensive poll of the global environmental community to gauge where we stand at this particular moment in history, the bottom line would be that our efforts to protect nature have not been up to the challenges posed by incessant human demands on the planet. A similar poll of cultural heritage practitioners would produce like results; anthropologists and linguists are deeply concerned about declining cultural diversity; site managers, about increasing physical impacts on landscapes and buildings; museum professionals, about flagging funding for the continued conservation of objects under their care; and so on. In short, there is widespread agreement among all the disciplines concerned with natural and cultural heritage that whatever it is we are doing, it is not enough. Because natural and cultural heritage conservation has proceeded, and continues to proceed, on mostly separate tracks with little interaction, one can reasonably conclude that applying an interdisciplinary approach might produce better results, and therefore should be tried.

The second line of reasoning is more direct, and comes from accumulating evidence that there are many inherent linkages between nature and culture that reveal themselves fully only under interdisciplinary scrutiny. The evidence is coalescing around the concept of 'biocultural diversity', which is defined as the sum total of the world's differences, no matter their origin. It includes biological diversity at all its levels, from genes to populations to species to ecosystems; cultural diversity (including linguistic diversity) in all its manifestations, ranging from individual ideas to entire cultures; abiotic diversity, or the differences found in the non-living components of Earth, such as geological features, landforms, climatic and chemical cycling systems, and so forth; and, importantly, the interactions among all of these.[13]

> On a global scale, the primary importance of biocultural diversity is that it is the fundamental expression of the variety upon which all life is founded. Conceptually, biocultural diversity bridges the divide between disciplines in the social sciences [and humanities] that focus on human creativity and behavior, and those in the natural sciences that focus on the evolutionary fecundity of the non-human world. The result is a more integrated view of the patterns that characterize life on Earth.[14]

The concept of biocultural diversity arose in the 1990s when studies first demonstrated a strong overlap between areas with high biological diversity and those with high linguistic diversity, and parallels began to be drawn between the extinction of species and languages.[15] Since then, biocultural diversity research has begun to mature into what Margaret A. Somerville and David J. Rapport have identified as a 'transdisciplinary field'—one that not only spans different realms of inquiry but also links theory, practice, policy, and ethics.[16]

As one example of how this line of research can have practical application to heritage conservation, the NGO Terralingua has developed a global Index of Biocultural Diversity (IBCD), which uses the number of languages, religions, ethnic groups, bird species, and mammal species to calculate country-level biocultural diversity scores for all the nations of the world. The IBCD uses statistical analyses of these indicators to create three complementary measures of biocultural diversity: one showing the 'raw' amount of a country's diversity (IBCD-RICH), another which adjusts for differences among countries in terms of their size (IBCD-AREA), and a third which adjusts for differences in their population (IBCD-POP). A major result of this work is its identification of three 'core areas' of biocultural diversity: the Amazon Basin, Central Africa, and Indomalaysia/Melanesia (see Figure 1). These are the world's richest areas in terms of overlapping biological and cultural diversity; from a conservation standpoint, they ought to be among the highest priority places for programmatic and external donor investment.

One way such investment might be targeted is by extending the concept of *biodiversity hotspots* to include *biocultural diversity hotspots*. A major thrust in biodiversity conservation today (albeit not without controversy), biodiversity hotspots are areas that hold exceptionally high levels of the planet's endemic plant and terrestrial

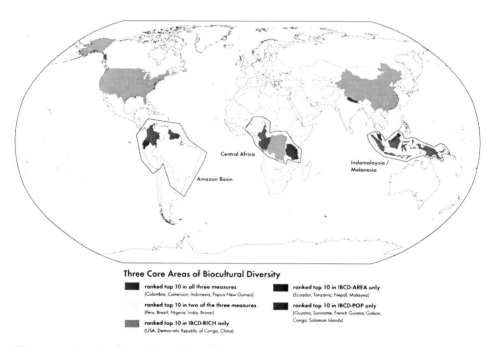

**Three Core Areas of Biocultural Diversity**

▪ ranked top 10 in all three measures
(Colombia, Cameroon, Indonesia, Papua New Guinea)

▪ ranked top 10 in two of the three measures
(Peru, Brazil, Nigeria, India, Brunei)

▪ ranked top 10 in IBCD-RICH only
(USA, Democratic Republic of Congo, China)

▪ ranked top 10 in IBCD-AREA only
(Ecuador, Tanzania, Nepal, Malaysia)

▪ ranked top 10 in IBCD-POP only
(Guyana, Suriname, French Guiana, Gabon, Congo, Solomon Islands)

**Figure 1**   Results from the Index of Biocultural Diversity point to three 'core areas' (outlined here) of biocultural diversity which contain at least one country that ranks highly in all three of the index's measures. These core areas are the Amazon Basin, Central Africa, and Indomalaysia/Melanesia. A subsidiary area of important biocultural diversity runs from India through Nepal to China (*Source*: map based on data summarised in Loh and Harmon, 'A Global Index of Biocultural Diversity.')

vertebrate species and which also are losing large percentages of their natural habitat. The hotspots concept has been taken up by the NGO Conservation International and reworked several times; today, 34 hotspots have been identified.[17] At least 11 of these fall within the 'core areas' of biocultural diversity mapped in Figure 1.[18] Urgent action is being advocated on behalf of threatened biodiversity within the hotspots, but in these 11 overlapping areas of high biocultural diversity, urgent action is also needed to stem the forces of globalisation that are driving language extinctions and other declines in cultural diversity. It seems an option worth exploring to try to tackle the two problems in concert, since the forces imperilling biological and cultural diversity are often one and the same.

Another example of a biocultural approach to protected area conservation is the work of IUCN on the biodiversity values of sacred natural sites. The task force on Cultural and Spiritual Values of Protected Areas, which is part of IUCN's World Commission on Protected Areas, has launched two projects on this topic. One, funded by the Global Environment Facility and carried out in partnership with the Rigoberta Menchú Tum Foundation, is called 'Conservation of Biodiversity-rich Sacred Natural Sites of Indigenous and Traditional Peoples'. The goal is to support the conservation and sustainable use of biodiversity found at these sites, focusing on five regions (Meso-America, South America, South Asia, East Africa, and West Africa). The specific objectives of the project are to increase awareness of sacred natural sites, bolster legal protections for them, strengthen the ability of traditional stewards to care for them, and share skills and field techniques that enhance their protection. A second, parallel project, called the Delos Initiative, encompasses sites in Europe, North America, Japan, and Australia/New Zealand. It aims to 'identify the pertinence and meaning of sacred natural sites found in the developed world, and to investigate whether and how spiritual values can contribute to the conservation and wise use of significant natural areas in this part of the world'. The focus is on sites with high biodiversity.[19] The IBCD and the work on sacred natural sites point the way towards integrated heritage conservation and the unique benefits that such an approach can offer. This paper concludes with four short ideas for making integrated heritage conservation a reality.

### Make a Sustained Effort to Identify Shared Ethics

It is not difficult to see that an ethic shared among nature conservationists, social scientists, and cultural advocates (both indigenous and others) will revolve around the paramount importance of diversity and the processes that create it. In my book *In Light of Our Differences: How Diversity in Nature and Culture Makes Us Human*, I make an extended case for diversity in nature and culture being the pre-eminent fact of existence. The book contends that diversity is the means by which our consciousness functions, and so it follows that if our consciousness is what makes us uniquely human, then diversity makes us human. This leads to the premise that we have a moral responsibility to maintain the diversity of the world's natural and culture heritage—together.[20]

This sort of reasoning has the potential to appeal to both nature conservationists and cultural heritage practitioners. The trouble is that few in these fields have experience in articulating the moral basis of their work. Ethical challenges invariably draw the most, and most passionate, interest from people working to protect natural and cultural heritage, but there is almost never any explicit discussion in the direct terms of ethics—it simply floats like ether in the background. When conflicting opinions arise, as they invariably do, they get tossed back and forth without there being any means of making the underlying assumptions explicit. So we go continually round and round without any hope of clarification, let alone reconciliation, of wide-ranging opinions.

One way out of this vicious circle is to provide opportunities to work with trained ethicists who can help people make their moral reasoning explicit. By aiding heritage workers to pinpoint their underlying moral values and see how those bear upon their work, ethicists can help them transcend the boundaries of their own specialisation: to help them, metaphorically speaking, first find, and then climb over, the stiles that exist along the fence lines of their field. For example, the Ethics Specialist Group of IUCN is exploring ways to train conservation professionals to understand the ethical underpinnings of their work as part of developing a code of ethics for biodiversity conservation. Likewise, every major cultural resource discipline has a code of ethics, while UNESCO, the Ename Center, and others have developed codes of applied ethics for groups such as dealers in cultural property, heritage site managers, heritage tourism operators, and so forth. What is lacking is any cross-fertilisation among these efforts. Yet there are many points of congruence, and much to be gained on both a theoretical and practical level by sharing experiences and working together to find common ground in what motivates people. In the end, this work may lead us to the following conclusion, expressed by leaders of the Ethics Specialist Group:

> It is impossible to separate the ethical obligations humans have for good and responsible relationships to the variety of life forms on Earth from the ethical obligations humans have for other humans. How humans treat one another affects their capacity to take responsibility for nature; indeed, in most cases the responsibilities in question *are responsibilities of and for biocultural systems*, integrations of human and non-human ways of life.[21]

## Search for Common Terms of Reference

To identify shared ethics, one must get people speaking at least a few words of a mutually intelligible language. These common terms of reference are not as scarce as might be supposed. The concept of *integrity*, and its close cousin *authenticity*, have real promise here. Ecological integrity is one of the four principles behind the Earth Charter, and when one compares it with the Nara Document on Authenticity (which builds upon and extends the Venice Charter), some striking resemblances jump out, despite their differences in tone. Both emphasise diversity as an irreplaceable source of spiritual and intellectual richness, both affirm that heritage protection is a universal responsibility, both emphasise respect as the basis for moral considerability.[22] Examination of other international instruments, such as the Universal Declaration of Human Rights, the draft UN Declaration on the Rights of Indigenous Peoples, the UNESCO Universal Declaration on Cultural Diversity, the Convention on Biological Diversity (especially its Section 8), and so on, would reveal similar commonalities.

Another possible common ground surrounds the concept of *restoration*. There is a rich practical literature on ecological restoration, and of course restoration of artistic and other cultural objects is a highly developed field. Has there been any attempt to apply the restoration theory of Cesare Brandi to the philosophical quandaries of nature restoration?[23] Or vice versa? Such efforts would make for a fascinating exercise.

## Get Serious about Interdisciplinarity

If the much-sought-after interdisciplinarity is to be really achieved, flagship journals with a stake in the enterprise, such as this one, *Conservation Biology*, etc., will have to make a commitment to publishing the fruits of multidisciplinary research. Paradoxically, the only way interdisciplinary conservation research will begin to have an impact is if it is published in specialist journals. This could well entail major changes in editorial policies. One might envisage, for example, leading journals entering into formal agreements simultaneously to publish key interdisciplinary articles so that the best of this work is disseminated at the same time to a wide range of specialist audiences. While there are conservation-oriented interdisciplinary research centres at several academic institutions, and while within the separate spheres of natural and cultural heritage disciplines there are prominent centres (such as ICCROM) that promote multidisciplinary work within those spheres, effectively to launch the enterprise a formal Integrated Heritage Conservation Research Consortium, involving professional societies, universities, government agencies, NGOs, and private companies, may be required.

What, practically speaking, are the prerequisites for successful interdisciplinarity in natural and cultural heritage conservation? Is it really essential for biologists who wish to make observations about human beings to know the anthropological literature thoroughly? No, but it is essential that they are willing firmly to grasp the main outlines of the discipline and its key points—to know, for example, why the differences between Iroquois cousin terminology and other kinship systems are important. What of anthropologists who wish to go deeper than dismissive judgements of 'environmental determinism' in an understanding of the interactions of biology and culture? Must they understand every nuance of evolutionary biology and the complexities of ecosystem function? Again, no, but they should be prepared to understand how the basic mechanisms of evolution function and how ecosystems operate.

Note the phrases 'are willing firmly to grasp' and 'should be prepared to understand'. There are no such things anymore as Renaissance men (and women)—if there ever were. No one can be competent in one's own specialty and be intelligently conversant, let alone expert, in all the disciplines that must be brought to bear on integrated heritage conservation. Rather, one must be willing and ready to absorb the main lines of complementary fields as part of an interdisciplinary team. The only way interdisciplinary

on can be made to work is as part of a *facilitated team* focused on problem
ence, the real prerequisites for successful interdisciplinarity are: (1) the pres-
trained and sensitive mediator/facilitator who is skilled in finding common-
lisparate points of view, and in getting people to recognise them and then work
together on that basis, and (2) the presence of workers who are sufficiently open-
minded to want to participate in such a facilitated process, and intellectually nimble
enough to contribute to its success.

### Focus on Practical Outcomes rather than on Differences in How the World is Understood

The point of integrated natural and cultural heritage conservation is not to try to stan-
dardise or homogenise fundamentally different ways of knowing the world. For
instance, it is an ineluctable fact that biologists trained in the Western scientific tradition will never put much credence in indigenous creation stories that explain natural
phenomena through supernatural forces. At the same time, we should not be surprised
when indigenous people, and others, decry the lack of spirituality of scientists, of what
they perceive to be soulless materialism and misguided reductionist thinking. These are
examples of basic epistemological disagreements, differences in what constitutes
knowledge and the legitimate ways to acquire it. There are certain 'hard facts' about the
world that are universal, not relative, and can be grasped by anyone regardless of their
cultural background, academic training (or lack thereof), individual talents, or
personal predilections. Beyond this, though, knowledge is not monolithic; so we have
to draw a sharp distinction between the agreed core of knowledge and the multifarious
differences in how the *significance* of that knowledge is interpreted.[24]

What is ironic is that people on both sides of the nature/culture divide, people who
share a commitment to the value of diversity, fail to recognise that such epistemological
differences are themselves instances of diversity, and that this epistemological diversity
is itself valuable. The whole argument over the role of culture in nature conservation
(and vice versa) arises from the tension issuing directly from epistemological diversity.
This tension has the potential to create opportunities for collaboration, rather than
blocking it. When it comes to conserving the world's natural and cultural heritage, the
way forward is to build on shared ethics, using common terms of reference, in a facilitated, interdisciplinary team whose participants focus on working together to achieve
progress on the ground rather than arguing about whose interpretation of the world is
'right'. This is a tall order, so integrated heritage conservation, like traditional nature
conservation, will still be a tough business. But in a world that is increasingly aware of
its own plurality, and of the mounting threats to its natural and cultural heritage, an
integrated approach has the promise to be more effective than what has gone before.

### Acknowledgements

I am indebted to Thymio Papayannis and Peter Howard for their capable and collegial
suggestions on an early draft of this paper.

# Notes

[1]   For more on this connection, see Harmon, 'Biodiversity and the Sacred'.

[2]   Wilson started the debate in the (in)famous concluding chapter of his encyclopedic overview, *Sociobiology: The New Synthesis* (1975). He provides a fascinating recollection of the unfolding controversy in his 1994 memoir, *Naturalist.*

[3]   Wilson, *Sociobiology*, 547.

[4]   Williams, 'Genes, Mind, and Culture', 29.

[5]   IUCN, UNEP, and WWF, *World Conservation Strategy.*

[6]   The proceedings were published in McNeely and Miller, *National Parks, Conservation, and Development.*

[7]   Paraphrasing West and Brockington, 'An Anthropological Perspective'.

[8]   Van Schaik and Kramer, 'Toward a New Protection Paradigm'.

[9]   See Wilshusen et al., 'Reinventing the Square Wheel'; Brechin et al., 'Beyond the Square Wheel'; and idem, *Contested Nature*, for policy reviews of the resurgent protection paradigm from a sociological perspective. For 'back to the barricades' and 'fortress conservation', see Hutton et al., 'Back to the Barriers?' and Christensen, 'Aux Barricades!'

[10]  Fewer still allow themselves to lapse into a sterile misanthropy, such as that characterised by rhetoric which likens people to a kind of cancer, metastasising and fatally infecting the Earth.

[11]  Meffe et al., 'Conservation Biology at Twenty', 596.

[12]  West and Brockington, 'An Anthropological Perspective', 611. See also Brosius, 'Common Ground'.

[13]  The definition is based largely on Loh and Harmon, 'A Global Index of Biocultural Diversity', 231–32; the clause about abiotic diversity is new here and my responsibility alone, having been inspired by the discussion in Gray, *Geodiversity*, 1–11.

[14]  Loh and Harmon, 'A Global Index of Biocultural Diversity', 232.

[15]  Another parallel between cultural and natural heritage, for which I am indebted to Peter Howard, is that the tendency to protect more and more examples of vernacular architecture and human settlements, rather than just superlative buildings, is mirrored by the drive to protect representative examples of all species lineages, rather than just the most charismatic species.

[16]  Somerville and Rapport, *Transdisciplinarity*, cited in Maffi, 'Linguistic, Cultural, and Biological Diversity', 612–13. Maffi's review article is the best current summary of the field of biocultural diversity research.

[17]  See the Conservation International website, 'Biodiversity Hotspots' [accessed 31 August 2006], available from http://www.biodiversityhotspots.org

[18]  The number would be higher if the India–Nepal–China axis of important biocultural diversity, referred to in the caption to Figure 1, were also included.

[19]  For the 'Conservation of Biodiversity-rich Sacred Natural Sites of Indigenous and Traditional Peoples' project, see G. Oviedo, S. Jeanrenaud and M. Otegui, 'Protecting Sacred Natural Sites of Indigenous and Traditional Peoples', unpublished report for IUCN, Gland, Switzerland [accessed 31 August 2006], available from http://www.iucn.org/themes/spg/Files/sacred%20sites/protecting-sacred-natural-sites-indigenous.pdf. For the Delos Initiative, see http://www.med-ina.org/delos/basic.htm

[20]  Harmon, *In Light of Our Differences, passim.* See pp. ix–xiv for a summary.

[21]  R. Engel, B. Mackey and K. Bosselmann, 'Toward a Code of Ethics for Biodiversity Conservation', unpublished draft proposal from the IUCN Commission on Environmental Law's Ethics Specialist Group, April 2006, 4.

[22]  The Earth Charter Initiative, 'The Earth Charter' [accessed 31 August 2006], available from http://www.earthcharter.org/files/charter/charter.pdf. For the Nara Document, see Larsen, *Nara Conference on Authenticity*, xxi–xxv.

[23]  Cf. Brandi, *Theory of Restoration*, and Gobster and Hull, *Restoring Nature.*

[24]　For an extended discussion of these points, see Kekes, *Pluralism in Philosophy*, esp. 25–46. See also the discussion of polythetic classification and polythetic morality in Harmon, *In Light of Our Differences*, 97–117, 135–39, 153–60.

## References

Brandi, C. *Theory of Restoration.* Translated by C. Rockwell. Florence: Nardini Editore, for the Istituto Centrale per il Restauro, 2005.

Brechin, S. R., P. R. Wilshusen, C. L. Fortwangler and P. C. West. 'Beyond the Square Wheel: Toward a More Comprehensive Understanding of Biodiversity Conservation as Social and Political Process'. *Society and Natural Resources* 15 (2002): 41–64.

———, eds. *Contested Nature: Promoting International Biodiversity with Social Justice in the Twenty-first Century.* Albany: State University of New York Press, 2003.

Brosius, J. P. 'Common Ground between Anthropology and Conservation Biology'. *Conservation Biology* 20, no. 3 (2006): 683–85.

Christensen, J. 'Aux Barricades!' *Conservation in Practice* 7, no. 2 (2006): 48.

Gobster, P. H. and R. B. Hull, eds. *Restoring Nature: Perspectives from the Social Sciences and Humanities.* Washington, DC: Island Press, 2000.

Gray, M. *Geodiversity: Valuing and Conserving Abiotic Nature.* Chichester: John Wiley, 2004.

Harmon, D. *In Light of Our Differences: How Diversity in Nature and Culture Makes Us Human.* Washington, DC: Smithsonian Institution Press, 2002.

———. 'Biodiversity and the Sacred: Some Insights for Preserving Cultural Diversity and Heritage'. *Museum International* 218 (2003): 63–69.

Hutton, J., W. M. Adams and J. C. Murombedzi. 'Back to the Barriers? Changing Narratives in Biodiversity Conservation'. *Forum for Development Studies* 2 (2005): 341–69.

IUCN (International Union for the Conservation of Nature and Natural Resources), UNEP (United Nations Environment Programme), and WWF (World Wildlife Fund). *World Conservation Strategy: Living Resource Conservation for Sustainable Development.* Gland: IUCN, UNEP, and WWF, 1980.

Kekes, J. *Pluralism in Philosophy: Changing the Subject.* Ithaca, NY and London: Cornell University Press, 2000.

Larsen, K. E., ed. *Nara Conference on Authenticity in Relation to the World Heritage Convention.* Proceedings of the conference, 1–6 November 1994, Nara, Japan. Trondheim: Tapir Publishers, for the UNESCO World Heritage Centre, ICCROM, and ICOMOS, 1995.

Loh, J. and D. Harmon. 'A Global Index of Biocultural Diversity'. *Ecological Indicators* 5 (2005): 231–41.

Maffi, L. 'Linguistic, Cultural, and Biological Diversity'. *Annual Review of Anthropology* 29 (2005): 599–617.

McNeely, J. A. and K. R. Miller. *National Parks, Conservation, and Development: The Role of Protected Areas in Sustaining Society.* Washington, DC: Smithsonian Institution Press, 1984.

Meffe, G. K., D. Ehrenfeld and R. F. Noss. 'Conservation Biology at Twenty'. *Conservation Biology* 20, no. 3 (2006): 595–96.

Somerville, M. A. and D. J. Rapport. *Transdisciplinarity: reCreating Integrated Knowledge.* Montreal: McGill-Queen's University Press, 2002.

Van Schaik, C. and R. Kramer. 'Toward a New Protection Paradigm'. In *Last Stand: Protected Areas and the Defense of Tropical Biodiversity*, edited by R. Kramer, C. van Schaik and J. Johnson. New York and Oxford: Oxford University Press, 1997.

West, P. and D. Brockington. 'An Anthropological Perspective on Some Unexpected Consequences of Protected Areas'. *Conservation Biology* 20, no. 3 (2006): 609–16.

Williams, T. R. 'Genes, Mind, and Culture: A Turning Point'. *Behavioral and Brain Sciences* 5 (1982): 29–30.

Wilshusen, P. R., S. R. Brechin, C. L. Fortwangler and P. C. West. 'Reinventing a Square Wheel: Critique of a Resurgent "Protection Paradigm" in International Biodiversity Conservation'. *Society and Natural Resources* 15 (2002): 17–40.

Wilson, E. O. *Sociobiology: The New Synthesis.* Cambridge, MA and London: Belknap Press of Harvard University Press, 1975.

———. *Naturalist.* Washington, DC: Island Press, 1994.

# Inspiration, Enchantment and a Sense of Wonder … Can a New Paradigm in Education Bring Nature and Culture Together Again?

Alan Dyer

It is the Child that sees the primordial Secret in Nature
and it is the child in ourselves that we return to.
The child within us is simple and daring enough to live the Secret.

> Chuang Tzu

Somewhere in almost every convention, declaration or strategy on nature, the environment, biodiversity or cultural diversity there will be a section enthusiastically espousing how education will be one of the major progenitors towards the successful implementation of the report. Education is undoubtedly at the heart of culture—any culture, ancient or modern—it is how we have shared our experience, 'taught' each other about our world and how to survive in it, enjoy it or exploit it. John Dewey's definition, '*Education is a social process. Education is growth. Education is not a preparation for life; education is life itself*',[1] underlines the fact that every one of us has been through an education system of some sort, formal or non-formal, and will have vivid memories and stories about our experiences. Those experiences could have been painful, dull or inspirational. Perhaps the reason you are reading this article now is because of an inspirational teacher—or the need to escape a boring, tyrannical tutor. Famous writers, thinkers and academics often tell of the meeting with a teacher, elder, university tutor, wild woman or wilderness man that sent them on their path to fame.

David Orr points out that the transition to a sustainable society is led by governments, corporations and individuals and that: 'the one thing these have in common are people who were educated in our public schools, colleges and universities'. He goes on: 'We may infer from the mismanagement of the environment throughout the century that most emerged from their association with these various educational institutions as ecological illiterates, with little knowledge of how their subsequent actions would disrupt the earth.'[2] Almost 15 years after Orr wrote that, there is little evidence that the levels of ecological literacy or sustainability literacy are much better in the majority of our educational institutions, or amongst our political or corporate leaders. It is a sobering thought that many of those responsible for wars, weapons development, cultural genocide, species extinction, GM crops, polluting industries and high levels of consumption are graduates from our universities! Whose fault is this? Schools, universities, teachers, quality assurance inspectors, parents, governments? Surely the fault lies at the feet of all of us; we have not always built on our rich personal childhood experiences—that close contact with nature and culture through first-hand experience—but we have all too often allowed fear to take away the magic of childhood and the joys of college and university life. And our entire culture is suffering as a consequence.

Clearly, that is not a universal picture; there are individual institutions in all sectors, across many nations that are leading the way forward—but it is not an easy path to tread. It would seem that there are very few free democratic states across the world that are not constantly revising, reconsidering or radically changing their education policies to cope with 'progress'. There are curricula across all sectors which now contain more elements of cultural heritage, ecology and sustainability—particularly in science, geography, citizenship and religious education—but somehow the holistic energy that turns facts into feelings, professional development into corporate responsibility and understanding into personal action is not there in sufficient strength. There is a distinct lack of magic and enchantment (this does not refer to the supernatural type of magic, simply those powerful 'awe and wonder' experiences that drew us into our subject— the 'wow' factor) coupled with a lack of holism that makes the overall experience memorable and life enhancing. As our world becomes more complex and sophisticated there is a tendency to add more to an already overcrowded curriculum. As we take more pupils and students into our institutions there is considerable pressure on space and staff time. Very often the first elements to be cut when more needs to be added are the experiential field trips, the out-of-class practicals, the time to play, the space for reflection, the encouragement of interdisciplinarity, the challenge to experience other working practices or even time for 'hanging out with friends'!

No matter where in the world one travels young children can be seen (roughly in the 3–12 age range) thoroughly involved in learning about the Earth, its life and human culture. If they get dirty they hardly notice, if they fall then any tears are short lived and if they have little in the way of 'things' then they have an imagination that is staggering. School may or may not be a central part of this process but for the majority of these children life is a 'hands-on' journey of discovery, laughter and adventure. The 'play' element in education is now universally regarded as centrally important to a child's development and builds a social context into which facts, knowledge and understanding have a fertile space to grow. Elementary, or primary, schools I have visited high in the mountains of Patagonia, in tropical West Africa, inner city-Taipei, rural Devon in England or small-town California all have programmes in Earth care and people care. The majority of children and teachers I met there seemed tuned into the need for recycling, energy saving, promoting biodiversity, understanding their immediate surroundings, a sense of beauty and most importantly, expressing their thoughts and findings in a dynamic way.

Many schools, centres and heritage sites use or adapt programmes that originated in North America in the 1970s and 1980s, such as Joseph Cornell's *Sharing Nature*,[3] Steve Van Matre's *Earthwalks*[4] and *Sunship Earth*[5] and Tom Brown's *Field Guides*.[6] These all involve direct contact with the natural world, use and develop all the senses, have a storyline that is both fun and exciting and also teach basic scientific concepts. Children are thoroughly involved at all times, are encouraged to express thoughts and feelings, and the programmes hopefully lead to some environmental action on their part. This approach was often criticised for being too 'touchy-feely', being too focused on games and not academically rigorous. However, look more closely and you will find excellent examples of discovery learning, programmes that have a carefully crafted rationale, are intricately planned, well prepared, fully resourced and with measurable learning outcomes. Far from simply being unscientific games, students coming out of the *Sunship Earth* programme, for example, will have *experienced* in a thoroughly memorable way the following concepts in (to quote Piaget) a 'concrete operational' way: Energy Flow, Cycling, Diversity, Community, Interrelationships, Change and Adaptation. And they will have remembered them through the mnemonic EC-DC-IC-A. These are seven hugely important concepts which underpin a basic understanding of how the natural world operates—appropriate from elementary school to university-level courses. Cornell's 'Flow Learning'[7] puts a range of simple environmental games into a powerful sequence built on sound educational principles. Children and students have a truly memorable experience, often at an important heritage site, and teachers/ leaders have a context in which to develop further work in many areas of the curriculum. Bringing concrete understanding to abstract scientific concepts through memorable experiences in an inspiring environment is a very powerful educative tool which too few programmes achieve.

So, although childhood may be seen to be under pressure from many angles in the modern world, the early years of schooling are potentially well served. What happens when these children become teenagers and go on to high school? Are there comparable programmes that take account of their growing independence, their raging hormones, irrational peer pressure, stressful examinations and a need to express themselves differently? Generally, in the state sector, there are not! Unless these schools have a programme such as the international baccalaureate, these holistic programmes do not exist in mainstream education. There are examples of excellent private schools and colleges which provide a superbly balanced curriculum—but they are accessible only to a minority.

Children, in the UK at least, are expected to choose their subject specialisms at a very early age, and be constantly tested on their progress. There are specialist art, music, science, drama or sports academies—but how much do they take account of ecological literacy? There are programmes such as '*Sunship 3*' from the Institute for Earth Education[8] or *Rediscovery*[9] that are interactive, challenging, involve the students directly and express good science through the expressive arts and a powerful environmental or cultural ethic. *Rediscovery* was an attempt to bring the 'elders' of NW Pacific coast indigenous populations back in touch with their young people to learn about their language, traditions and culture. The programmes involve camps, expeditions, crafts and traditional medicine, and are carefully designed to appeal to teenagers and pull them, rather than push them, into the scheme.

But these programmes are offered only in very few environmental centres, holiday courses and summer camps. Could not mainstream education learn from these programmes? Rather than compartmentalise these young minds into rigid subject areas and specialisms should we not be taking them on a wider path? How do we teach the science specialist the joys of art and the artist the intricacies of the web of life? Fritjof Capra argues that 'because the study of patterns requires visualising and mapping, every time that the study of pattern has been in the forefront, artists have contributed significantly to the advancement of science'. He quotes Leonardo da Vinci and Goethe as examples from the past and explains how modern systems thinking can be applied to integrate fragmented academic disciplines. He concludes:

> It is no exaggeration to say that the survival of humanity will depend on our ability in the coming decades to understand these principles of ecology and to live accordingly. Nature demonstrates that sustainable systems are possible. The best of modern science is teaching us to recognise the processes by which these systems maintain themselves. It is up to us to learn to apply these principles and to create systems of education through which coming generations can learn the principles and learn to design societies that honour and complement them.[10]

When our high school children go on to college or university do they have a sound understanding of the principles of ecology or the social systems that allow human cultures to progress? Do they leave university with a better understanding of social justice, equity or sustainability? A narrow view can sometimes be all too apparent and as a result we hear from disgruntled employers that they are astounded by the lack of breadth and initiative their new graduates demonstrate.

So, the message is clear, we need more 'enchanters'—or put into educational jargon, we should call for the development of more interactive pedagogies and a new paradigm of transformative, rather than transmissive, education that encourages direct experience and engagement across all sectors of formal and non-formal education. Is that not central to the role of all of us concerned with the natural world, cultural understanding and heritage studies?

Whatever one's interpretation of the evidence for climate change, habitat loss, threats to biodiversity and ecological stability we can now postulate that we are witnessing a period of rapid change due to human activity which will require us to take immediate action. The general level of awareness of the problems and possibilities has been raised to an unprecedented level through a sophisticated and global media. Those media are asking for solutions and encouraging the public to demand answers and action. There is a strong cynicism of politicians and the political process so the public are often looking to us, the academics, teachers and experts. So, yet again, education is heralded as a major tool for precipitating awareness into action and delivering the practical, moral, ethical and behavioural changes necessary to avert global disaster. But what real changes are being effected in our schools, universities and heritage centres which truly engage people with global issues and local action? It is very easy to say that 'they' should do something, but if we wait for 'them' to take action we will make little progress. It is up to us, as individuals, now, to use our experience, skills and wisdom to lead by example.

Environmental education, since its inception in the early 1960s, has succeeded in raising awareness of issues and problems, but has largely failed to deliver comprehensive action. Environmental education has evolved into Education for Sustainable Development (ESD)—indeed, we are in the UN decade of ESD (2005–2014).[11] ESD is a hotly contested area, particularly regarding the definitions of, or inclusion of, the word 'development'. However, we should avoid the debate as to whether the title should be Learning for Sustainability, Education for a Sustainable Future, Sustainability Literacy, etc. and concentrate on the fact that this change has significantly shifted the focus from natural science and purely environmental issues to a wider debate of social justice, equity, inter- and intra-generational interaction and the elimination of poverty. That has to be a positive change, but has the education system taken account of the rapid changes in the context in which we deliver these issues? The threats of rapid climate change, environmental degradation and social instability are now more than conjecture and the 'population time bomb' predicted in the 1970s is with us. Has there been an adequate response from education systems around the world?

Will the UNESCO decade for ESD deliver enough of a response to all sectors to ensure a co-ordinated, sequential and cumulative education both for our young people as they move through the formal system from nursery to university, and Continuing Professional Development (CPD) for the professionals who have responsibility for action now?

In his Schumacher Briefing Paper entitled *Sustainable Education*, Stephen Sterling describes the value of 'whole systems thinking' as a way to '… go beyond the dominant forms of thinking which are analytic, linear and reductionist. He helpfully says '… it offers a way of making holistic thinking understandable, accessible and practicable'. In outlining the shift from mechanistic to ecological thinking he asserts that

> It identifies three interrelated dimensions of paradigm—perceptual, conceptual and practical—which describe human experience and knowing at any level—personal, group or whole societies. In the dominant paradigm, there tends to be dis-integration within and between these dimensions. For example, in the Western tradition, intellectual knowledge (conceptual dimension or 'eidos') has primacy, to the extent that other forms of knowledge such as 'intuitive knowing' (perceptual dimension or 'ethos') or 'practical knowing' (practice dimension or 'praxis') are often regarded as having less value.[12]

Consider the practical application of the dominant paradigm of mainstream education today, at all levels from pre-school to PhD, and it often seems far from the spirit of the root of the word—to educe or to draw out. Rather, a more appropriate root might be *Sarcina*, a pack, a burden, a load, which may better describe the prescriptive cramming, testing and assessment to which we incessantly subject our students. The response to a rallying cry of 'Education, Education, Education',[13] 10 years on, has been a weary Plod, Plod, Plod! The fact that England has a 'curriculum' of sorts for the 0–3 age range (*Birth to Three Matters*)[14] must say something of the intellectual cage our teachers and students are in. Where is the magic, the unforgettable experience that had us on the edge of our chairs or up to our ears in mud? Where is the *ethos* and above all, the *praxis* that Stephen Sterling is calling for? Add to this the growth of a litigant culture, a preoccupation with risk assessments and unrealistically sterile conditions and the 'inspirational experience' we wish for our pupils and students vanishes into mediocrity and boredom. As I write this, a major British broadsheet newspaper has the headline, on the front page, 'Junk Culture is Poisoning our Children' and refers to a letter from a hundred academics, writers and children's experts who '… blame fast food, computer games and competitive schooling for a rise in depression'.[15] If we truly wish to bring Nature and Culture together we need to provide people of all ages with experiences that not only inspire and enchant but also engage at a level of personal understanding and relevance. Education is clearly not just about children and schools—this paper will use examples from children's education but my experience as a Teaching Fellow in a university involves working with a wide range of people including children, students, teachers, rangers, heritage interpretation specialists, fellow academics and the general public. Many of the arguments relating to schools cut across sectors and argue for a holistic response rather than a sector- or discipline-related response.

In his review of Jerome Bruner's recent book *The Culture of Education*[16] Scott London discusses '… the building of a cultural psychology that takes proper account of the historical and social context of participants'. He postulates that there are 'essentially two ways by which we organize and manage our knowledge of the world: logical-scientific thinking, and narrative thinking. Schools traditionally favour the former and treat the narrative arts—song, drama, fiction, and theatre—as more "decoration" than necessity. "It is only in the narrative mode," Bruner points out, "that one can construct an identity and find a place in one's culture. Schools must cultivate it, nurture it, and cease taking it for granted.'"

Bruner goes on to explore an 'emerging thesis' in educational theory based on the concept of folk psychology, or folk pedagogy. This view holds that the way teachers instruct their students is determined to a great extent by the lay theories or implicit assumptions they have about how children learn. These intuitive theories, or folk pedagogies, are reflected in many of the common assumptions teachers have about children—that they are wilful and need correction, that they are innocent and must be protected from a dangerous or vulgar society, that they are empty vessels to be filled with knowledge that only adults can provide, that they are egocentric and in need of socialisation, etc. According to Bruner, once we realise that a teacher's conception of a learner shapes the instruction he or she provides, then equipping teachers with the best available theory of the child's mind becomes crucial.[17]

So how do we engage a wider public with a broader ecological thinking and ensure that *ethos*, *praxis* and, quite simply, first-hand sensory experience become part of their learning and everyday experience? David Orr argues for a hands-on experiential approach and quotes Aldo Leopold's 1941 statement 'that most Americans have no idea what a decent forest looks like. The only way to tell them is to show them.'[18] Orr then imagines what it would be like if we cannot show them a 'real' forest, just 'impoverished remnant forest or industrial forest where trees are grown like corn …'[19] Here then, we have a tangible reason for bringing together nature and culture. The need for conservation and wilderness preservation are becoming more and more pressing as urbanisation increases—by 2007 more than 50% of people will live in cities so we will have fulfilled long-held predictions and undoubtedly become an urban species. Our ecological footprint is leaving less and less space for wilderness, wildlife and even adequate access for outdoor recreational activities for all but the rich.

Richard Louve sees a rise of what he calls 'cultural autism'[20] as urban and suburban children display symptoms of 'atrophy of the senses' as they become more and more isolated from the natural world and an electronic technology takes away the excitement—and risk—of self-discovery and adventure.

In a recent issue of *Resurgence* magazine[21] I outlined how for many years I have taken Rachel Carson's *The Sense of Wonder* as the words that guided my teaching:

> If I had influence with the good fairy who is supposed to preside over the christening of all children, I should ask that her gift to each child in the world be a sense of wonder so indestructible that it would last throughout life, as an unfailing antidote against the boredom and disenchantments of later years, the sterile preoccupation with things that are artificial, the alienation from the sources of our strength.

> If a child is to keep alive his inborn sense of wonder without any such gifts from the fairies, he needs the companionship of at least one adult who can share it, rediscovering with him the joy, excitement and mystery of the world we live in. Parents often have a sense of inadequacy when confronted on the one hand with the eager, sensitive mind of a child and on the other with a world of complex physical nature, inhabited

by a life so various and unfamiliar that it seems hopeless to reduce it to order and knowledge ...

I sincerely believe that for the child, and for the parent seeking to guide him, it is not so important to know as to feel. If facts are the seeds that later produce knowledge and wisdom, then the emotions and the impressions of the senses are the fertile soil in which the seeds must grow ...[22]

Rachel's seminal essay—first published in July 1956 in *Woman's Home Companion* under the title 'Help Your Child to Wonder'—emphasises the need for adults to share their experience, at the child's level and with a childlike attitude: a sense of discovery, adventure and enjoyment.

Children the world over have a right to a childhood filled with beauty, joy, adventure and companionship. They will grow towards ecological literacy if the soil they are nurtured in is rich with experience, love and good examples. But they must have the freedom to discover things for themselves; they must not be wrapped in cotton wool and supervised every minute of the day. The only way they will learn social interaction, team building, self-confidence and self-esteem is through active participation. One parent recently commented: 'I want my child to take risks—but not to be at risk!' So this is a reminder that we must make huge efforts to share our experience as grown ups; we must devise new strategies for sharing our accumulated wisdom (remembering to fill out the risk-assessment form to ensure that the possibility of serious accidents is reduced). At a recent lecture given by a representative of the Royal Society for the Prevention of Accidents (ROSPA)[23] my attention was immediately grabbed by the speaker's opening remarks '... we hope children *will* have accidents'. He went on to explain that they want to prevent *serious* accidents, but minor accidents are often the only way children will learn about their capabilities and strengths. They should be able to make mistakes, and as a result there will be the inevitable grazed knees and elbows, they receive the odd cut or bruise—but hopefully in a generally safe environment.

How do we equip today's parents, teachers, leaders and interpretation professionals to meet the new challenges? Many of them have come through a home-life and education system which has discouraged adventure, discovery or an holistic, interdisciplinary approach to teaching and learning. Many have been exposed to a litigant culture that has fomented a deep fear of 'making mistakes' and being held personally responsible by an aggressive management system and an unsympathetic, sensation-seeking media. Teachers, lecturers and interpretation professionals have surely not changed in their ability or enthusiasm for innovation and excitement, nor has their talent as communicators and facilitators diminished in any way. What has diminished is the support from the meta-system that allows them to respond fully to the challenges of sustainability and a changing culture. They are working in a teaching and learning system that has effectively 'sold out' to a target-driven, hyper-competitive roller coaster leading to heightened anxiety and depression amongst children and students.

In the first instance teachers at all levels need to gain the confidence to be able to follow their intuition. To do this we must help them challenge the fear culture and make sure they are equipped with the skills to cope. This will involve being aware of current legislation, ensuring that health, safety and risk assessment regulations are known and adhered to, first aid and inclusion provisions are exemplary and all forms and paperwork are up to date and complete. Depressing—but essential. A good example for this is given by the Forest School movement. Born from a Danish example, this group has had great success in taking school children into the woods despite the fear culture around such activities. In its infancy the approach was criticised for being 'too safe, too sterile' but in fact these attributes have been the very ones that have built confidence and allowed the approach to flourish.

Secondly, they need the tools to be able to enchant. Our recent book, *Let Your Children Go Back to Nature*,[24] aims to provide the tools to enable adults, be they parents or teachers, to take on the role of guardian elders—or Enchanters, as we prefer to call them, for our aim is to re-enchant our land in the imagination of children: to give young adventurers the opportunity to enter the magical and perilous worlds of Nature, Landscape and Myth, and to become transformed by the experience. But we must consider the background to childhood: the archetypal innocent and potential hero. Edith Cobb writes:

> The study of the child in nature, culture and society reveals that there is a special period, the little-understood, prepubertal, halcyon middle age of childhood, approximately from five or six to eleven or twelve—between the strivings of animal infancy and the storms of adolescence—when the natural world is experienced in some highly evocative way, producing in the child a sense of some profound continuity with natural processes and presenting overt evidence of a biological basis of intuition.[25]

In archaic Greece, as in most traditional societies, it was the custom to subject young boys to a period of training, testing and questing in the wilderness. There are indications that this started when they were about nine, though Achilles began at six. The boys were taken from their mother's protective care to spend their time exploring and getting to know the countryside and its flora and fauna, playing hunting games, taking part in races and other sports, and fighting mock battles, with the training culminating in a challenging adventure. In Minoan Crete this process lasted two months, probably finishing at Midsummer.

Selected Greek girls of the same age were transformed into *arktoi* ('She-bears'), and they followed Artemis, goddess and Lady of the Wild Things, into the wild woods and hills for a similar form of initiation. This experience usually took place in forest surroundings where the children could learn to master their fears, develop new skills, and discover self-reliance. They faced a series of hazards and strange encounters prepared by their elders who would act as guides, guardians and challengers.

Throughout the ages in all traditional societies the latter would sit them around a fire at twilight and recount the myths, legends and wonder-tales of the tribe: stories in which a hero or a heroine enters a Dark Forest and overcomes fearsome foes—wily dragons, threatening ogres, malevolent witches, or evil knights. In the stories they learned that the forest is a place of challenge, discovery, transformation and shelter, which provides the brave and virtuous with unexpected help, often in the form of magical animals who reward kindness with rescue from dangers. This storytelling was an essential preliminary, offering warnings and helpful advice to the apprehensive novices about to embark on their first great adventure in life.

So exactly what are our children and students to learn as they embark on the adventure of formal education? How do we move them from awareness to action? Is it through learning facts, names, key concepts, skills, attitudes or far more complex and varied than that? Bill Scott[26] points out that in Germany, Education for Sustainable Development has gone down the line of competencies. Specifying not what is to be taught, but what is to be learned: 'A competence-oriented education ... asks: what problem-solving strategies, concepts, and abilities for social action should [pupils] have? ... [focusing on] ... motivations, and ... experiences ... will increase pupils' interest in the acquisition of skills.'

He elaborates the eight ESD competencies that are specified under the heading *Gestaltungskompetenz*—which means the specific capacity to act and solve problems:

- foresighted thinking;
- interdisciplinary work;
- trans-cultural understanding and co-operation;
- participatory skills;
- planning and implementation skills;
- empathy, compassion and solidarity;
- self-motivation and motivating others;
- distanced reflection on individual and cultural models.

Scott proposes the idea '... that those who possess such competence can help, through their active participation in society, to modify and shape the future, and to guide social, economic, technological, and ecological change. As you can imagine, these are elaborated upon in some detail, and seem to be applicable to both school and adult contexts.'

Education is more than simply teaching and learning, it is a complex mixture of almost infinite possibilities and needs the accumulated wisdom and experience of our complete cultural evolution in which to grow. It would seem appropriate to end with Rudolph Bahro's often-quoted phrase: 'When the forms of an old culture are dying, the new order is established by those who are not afraid to be insecure.'[27] If indeed a new culture is emerging, those establishing the new could do well to start with 'Education, Education, Education' and build this into 'Inspiration, Enchantment and Action!'

## Notes

[1] John Dewey's declaration concerning education. First published in *The School Journal* LIV, no. 3 (16 January 1897): 77–80.
[2] Orr, *Ecological Literacy*.
[3] Cornell, *Sharing Nature with Children*.
[4] Van Matre, *Earthwalks*.
[5] Van Matre, *Sunship Earth*.
[6] Brown, *Tom Brown's Field Guide to Wilderness*.

[7]   Cornell, *Sharing Nature with Children II*.
[8]   Van Matre, *Sunship 3*.
[9]   Henley, *Rediscovery*.
[10]  Capra, 'Speaking Nature's Language'.
[11]  UNESCO Decade of ESD (2005–2014) [accessed 3 October 2006], available from http://portal.unesco.org/education/en/ev.php-URL_ID=27234&URL_DO=DO_TOPIC&URL_SECTION=201.html
[12]  Sterling, *Sustainable Education*.
[13]  'Education, Education, Education' was one of the maxims that swept the UK Labour Party to power in 1997.
[14]  *Birth to Three Matters—A Framework to Support Children in their Early Years*, Sure Start, DfES, HM Government publications, November 2005 [accessed 3 October 2006]. Available from http://www.surestart.gov.uk/resources/childcareworkers/birthtothreematters/aboutthe-framework/
[15]  Ben Fenton, *The Daily Telegraph*, 12 September 2006 +editorial comment and letters.
[16]  Bruner, *The Culture of Education*.
[17]  Scott London [accessed 12 November 2006], available from http://www.scottlondon.com/reviews/bruner.html
[18]  Leopold, *The River of the Mother of God and Other Essays*.
[19]  Orr, *Earth in Mind*.
[20]  Louve, *Last Child in the Woods*.
[21]  Dyer, 'A Sense of Adventure'.
[22]  Carson, *The Sense of Wonder*.
[23]  Not being sure whether this is the official policy of the ROSPA speaker or a personal view, neither the speaker nor the conference are credited here.
[24]  Hodgson and Dyer, *Let Your Children Go Back to Nature*.
[25]  Cobb, *The Ecology of Imagination in Childhood*.
[26]  W. Scott, Keynote address at the UK launch of the UNESCO Decade for ESD, 13 December 2005, 'ESD: What Sort of Decade? What Sort of Learning?'
[27]  Bahro, *From Red to Green*.

## References

Bahro, R. *From Red to Green*. New York: Schocken, 1984.
Brown, T. *Tom Brown's Field Guide to Wilderness Survival*. New York: Berkley Books, 1983.
Bruner, J. *The Culture of Education*. Cambridge, MA: Harvard University Press, 1996.
Capra, F. 'Speaking Nature's Language: Principles for Sustainability'. In *Ecological Literacy: Educating Our Children for a Sustainable World*. San Francisco: Sierra Club Books, 2005.
Carson, R. *The Sense of Wonder*. New York: HarperCollins, 1998.
Cobb, E. *The Ecology of Imagination in Childhood*. New York: Columbia University Press, 1977.
Cornell, J. *Sharing Nature with Children*. Nevada City: Dawn Publications, 1978.
———. *Sharing Nature with Children II*. Nevada City: Dawn Publications, 1999.
Dyer, A. 'A Sense of Adventure'. *Resurgence* 226 (September/October 2004): 25–27.
Henley, T. *Rediscovery: Ancient Pathways, New Directions*. Vancouver: Western Canada Wilderness Committee, 1989.
Hodgson, J. and A. Dyer. *Let Your Children Go Back to Nature*. Milverton: Capall Bann, 2003.
Leopold, A. *The River of the Mother of God and Other Essays* by Aldo Leopold. Edited by S. Flader and J. B. Callicott. Madison: University of Wisconsin Press, 1991. First published 1941.
Louve, R. *Last Child in the Woods*. Chapel Hill, NC: Algonquin Books of Chapel Hill, 2005.
Orr, D. *Ecological Literacy: Education and the Transition to a Postmodern World*. Albany: State University of New York Press, 1992.

———. *Earth in Mind: On Education, Environment, and the Human Prospect.* Washington, DC: Island Press, 2004.

Sterling, S. *Sustainable Education: Re-visioning Learning and Change.* Totnes: Green Books/Schumacher Society, 2001.

Van Matre, S. *Sunship Earth.* Martinsville: American Camping Association, 1979.

———. *Earthwalks.* Warrenville: Acclimatization Experiences Institute, 1980.

———. *Sunship 3.* Warrenville: Acclimatization Experiences Institute, 1994.

# Biological Diversity and Cultural Diversity: The Heritage of Nature and Culture through the Looking Glass[1] of Multilateral Agreements

Peter Bridgewater, Salvatore Arico & John Scott

## The Apparent Dialectic between Biological and Cultural Diversity

At the outset we recognise that the planet is a cultural and biological kaleidoscope. In the 14 years since the United Nations Conference on Environment and Development

(UNCED), and 20 years since the term was first coined, significant advances have been made in the global management of biodiversity. But while environmental problems have become globalised, their potential management solutions have become more localised.

A growing body of evidence recognises the links between biological and cultural diversity and continues to explore the interface between these and other forms of diversity. In doing so, the role of indigenous peoples[2] both as custodians of biodiversity and proponents of cultural diversity is paramount in understanding the interconnectedness of these issues. But both of these diversities are part of the global heritage of humankind. Cultural heritage is obviously a deeply human phenomenon, but now we understand more clearly the role humans—and their heritage—have had, and continue to have on what is often called 'natural heritage'. In effect we believe this to be a false dichotomy, and that the two forms of heritage are but two sides of the same coin, and a reflection of the two diversities. For simplicity, in the following text we will speak only about diversity, but always with the heritage background outlined above.

Conservation of nature is at the heart of the cultures and values of indigenous peoples. For more than 300 million indigenous peoples, the Earth offers not only life but also the basis of their cultural and spiritual identities. Because their worldview holds that the Earth and its resources are inherited from the ancestors, and includes trans-generational obligations, the Earth and its resources are a sacred heritage. Indigenous peoples have sacred sites on their lands and waters—such as a child dreaming site for Aboriginal women where they believe the spirit of the baby enters their womb—or a totem dreaming place that is significant for the continuation of certain and sometimes sacred species.

That sacredness comes from the intergenerational obligation to pass the land on to future generations; including the obligation to use it in sustainable ways. Native Americans talk in terms of the seventh generation when making decisions about development. Traditional territories of indigenous peoples are clan estates and the extended family has obligations to that estate. Many indigenous peoples do not believe in land ownership in the Western sense, but instead see themselves as belonging to the land, and/or sea. Yet we need to identify a balance between what is often seen as indigenous imperialism (by urban people) versus an imposition of notions of pristine wilderness (by indigenous peoples)! In reality both views need to be accommodated in a framework which seeks to reconcile stewardship of natural and cultural heritage.

Global conventions, from the Convention on Biological Diversity (CBD) through the Convention to Combat Desertification (UNCCD), the Convention on International Trade in Endangered Species of Wild Fauna and Flora (CITES), the International Convention for the Regulation of Whaling (ICRW) and even the World Heritage Convention (WHC), among others, have tended to create a 'lowest common denominator' approach to resource management, which often ignores—or even militates against—aspects of cultural diversity, including adequate respect and understanding of indigenous peoples and their rights. The Commission on Sustainable Development and its priorities also contributes to this blandly globalising effect.

In many of the current discussions about environmental issues at a national and international level, people are not often treated or regarded as part of the biosphere; certainly not as part of biodiversity. Conservation of biodiversity is rooted in a subset of values drawn from broader cultural values, values that determine how we deal with nature. Broadly speaking, these values remain uncomfortably ambivalent as to the status of humans with respect to nature. Not surprisingly, conservation biologists and wildlife managers tend to focus on biological issues when addressing *reservation* of natural areas, but the achievement of *conservation* outcomes requires an understanding of people and their aspirations and an awareness of the political and social climate.[3]

Located in biologically diverse regions, indigenous peoples are pivotal to the whole equation of diversity. Take linguistic diversity (a characteristic of cultural diversity); in total, 4,635 ethno-linguistic (or cultural) groups representing 67% of an approximate world total of 6,867 ethno-linguistic groups could be termed 'indigenous'.[4] Considering indigenous peoples are less than 5% of the world's population, such statistics are evidence of both their great cultural diversity and their fragility.

Indigenous peoples have accumulated vast amounts of ecological knowledge in their long history of managing the environment. Such knowledge is embedded in their languages. When languages become extinct, associated traditional ecological knowledge is often lost along with unique and legitimate worldviews.

Yet linguistic diversity is not the only cultural diversity which may be regarded as heritage. Culinary practices and, more generally, food-based practices are increasingly looked at as a manifestation of how biological and cultural diversity are blended. As an example, a recent assessment of how the hundreds of genetic varieties of corn (*Zea mais*) have been promoted and conserved has shown the importance not only of breeding techniques and home gardens but also the culinary use of such varieties.

Linguistic diversity and biodiversity appear to be the result of a process of co-evolution.[5] Over time, human communities interact closely with their local environment; modifying it as they adapt to life in specific ecological niches, they acquire intimate and specialised knowledge of the environment and how to use and manage it for individual and group survival. Indigenous peoples may be seen as ecosystem peoples and their cultures are often cultures of habitat. We should note that cultural diversity and promotion of cultural heritage does *not* require that we tolerate practices that are against human rights. Indigenous peoples negotiating the Declaration on the rights of indigenous peoples have agreed to work within a human rights framework; and that they must live according to human rights norms.

## Scale and Diversity

One of the significant problems in any discussion of cultural and biological heritage is to maintain awareness of scale and of the existence of more than one scale. There are three basic and interactive elements of diversity: cultural, biological and place (spatial). The importance of all three elements should not be minimised, nor should one be allowed to dominate. Human identity is derived from the intellectual interpretations of the interactions of these elements. And this is where globalisation has most impact.

Prevailing values derived from the current beliefs of society can be influenced and shaped over time by information gathered scientifically, but at any given moment those values and beliefs may be more important in the shaping of public policy than the results of the latest scientific research. Cultural heritage also includes religious heritage, and spirituality can have effects beyond simply appreciating nature, through, *inter alia*, sacred forests and sacred groves.[6]

The approaches taken by multilateral environmental agreements to reconciling cultural and biological diversity are as different as the agreements themselves, and not always fully effective. The present review takes a broad view of some key agreements at the nexus of the two diversities, and proposes future action.

## UN Approaches: UNEP and UNESCO

At the occasion of the World Summit on Sustainable Development (WSSD, 2002), the United Nations Educational, Scientific and Cultural Organisation and Environment Programme (respectively, UNESCO and UNEP) convened a high-level round table on 'Cultural Diversity and Biodiversity for Sustainable Development'. At the round table, issues such as diversity and sustainable development, diversity in nature and culture and towards a culture of sustainable diversity were discussed.[7]

The degree of uncertainty surrounding the relationship between biological and cultural diversity, and the level of interest by participants in the round table, convinced UNESCO and UNEP to pursue joint work in this area. A 2003 UNEP Governing Council Resolution[8] on environment and cultural diversity referred to the importance of further examining this issue in co-operation with UNESCO, with particular attention to its implications for human well-being.

Within UNESCO's normative work, the interplay between cultural diversity and nature is well reflected in the 2003 Convention for the Safeguarding of the Intangible Cultural Heritage, the 2001 Convention on the Protection of the Underwater Cultural Heritage and the 1972 Convention concerning the Protection of the World Cultural and Natural Heritage, while the UNESCO Universal Declaration on Cultural Diversity specifically mentions the relation with biological diversity.[9]

Article 1 of the Declaration states:

> Culture takes diverse forms across time and space. This diversity is embodied in the uniqueness and plurality of the identities of the groups and societies making up humankind. As a source of exchange, innovation and creativity, cultural diversity is as necessary for humankind as biodiversity is for nature. In this sense, it is the common heritage of humanity and should be recognised and affirmed for the benefit of present and future generations.

The parallel with biodiversity is well illustrated: as much as biodiversity, cultural diversity contributes to the resilience of humankind.

In the case of human-dominated nature, which is the nature we face nowadays,[10] diversity of genes, species, ecosystems and landscapes allows us to develop a matrix of human activities while maintaining those benefits of nature that are important to human well-being. Diversity of culture in its different expressions contributes to the sustainability of human interactions and therefore provides an important contribution to the human element of sustainable development. A detailed review has been published on this issue by UNEP, following the publication of the Global Biodiversity Assessment.[11]

One of the main lines of action for implementing the Universal Declaration on Cultural Diversity refers to '[r]especting and protecting traditional knowledge, in particular that of indigenous peoples; recognising the contribution of traditional knowledge, particularly with regard to environmental protection and the management of natural resources, and fostering synergies between modern science and local knowledge.'

Within UNESCO, several arguments have been advanced to justify further work in this area:

- a need to demonstrate scientifically the importance of maintaining and promoting cultural diversity for human well-being;
- a need to enhance synergies between programmes of UNESCO and of sister UN bodies (e.g. UNEP) and agreements (e.g. CBD) with complementary mandates;
- a need to develop a conceptual model and a methodology to describe the relationship between the two expressions of diversity;
- a need to mainstream culture into the development agenda;
- a need to expand the ecosystem approach so as to broaden its scope and application beyond natural sciences;
- a need to provide empirical evidence illustrating the relationship between cultural and biological diversity.[12]

The recently adopted Declaration on the Rights of Indigenous Peoples[13] (which is regarded by indigenous peoples as the most up-to-date articulation of their rights), the Convention on the Protection and Promotion of the Diversity of Cultural Expressions (2005) and the Convention on Biological Diversity are most significant in exploring the interface between cultural and biological diversity.

The Declaration on the Rights of Indigenous Peoples (DRIPs) provides an insight into issues of relevance to indigenous peoples concerning cultural and biological diversity. Article 43 describes the Declaration as the minimal standard for the survival, dignity and well-being of the indigenous peoples of the world. References to culture and biodiversity are contained throughout the document (including the preamble and the substantive articles) but specifically Article 11 pertains to the right to practise, revitalise and maintain their cultures; Article 29 to the conservation and protection of the environment; and Article 31 to indigenous cultural heritage and traditional knowledge (including ownership of genetic resources, seeds, medicines, and knowledge of property of fauna and flora).

Also, the Convention on the Protection and Promotion of the Diversity of Cultural Expressions was the first legal instrument the international community possessed to raise cultural diversity to the rank of 'common heritage of humankind' inasmuch as biological diversity is a 'common concern of humankind'. In particular, Article 6 of the Convention (Principle of sustainable development) states that 'The protection, promotion and maintenance of cultural diversity are an essential requirement for sustainable development for the benefit of present and future generations.' This young convention follows the adoption of the UNESCO Universal Declaration on Cultural Diversity (2 November 2002) which states that: 'As a source of exchange, innovation and creativity, cultural diversity is as necessary for humankind as biodiversity is for nature' (Article 1).

The United Nations Millennium Declaration[14] refers to the diversity of nature ('the immeasurable riches provided to us by nature') and the diversity of humans ('[h]uman beings must respect one other, in all their diversity of belief, culture and language') as important values and principles that are essential in international relations in the 21st century for the purpose of achieving development in the new Millennium.

Physical expressions of cultural and biological diversity can be seen in sites recognised as biosphere reserves or World Heritage Sites under UNESCO's Man and the Biosphere (MAB) Programme[15] and World Heritage Convention (1972).[16]

### Resolving the Dialectic: Multilateral Environmental Agreements

*World Heritage Convention: Natural and Cultural in One Framework*

In 1992, the World Heritage Convention opened its list to encompass 'cultural landscapes' also, in addition to cultural sites and natural sites. For certain sites, the cultural aspects of their uniqueness are where landscapes have inspired and shaped specific cultural expression of unique value/nature; in others, it is the way culture has shaped the physical environment. The early cultural landscapes inscribed under the convention were termed associate, strongly emphasising the cultural links to the landscape.

For example, in 1993 Tongariro (New Zealand) became the first property to be inscribed on the World Heritage List under the revised criteria describing cultural landscapes. The mountain at the heart of the park has cultural and religious significance for the Maori people and symbolises the spiritual links between this community and its environment.[17] Later examples provided more explicit links between nature and culture. There is also a view that cultural landscapes are solely a product of the World Heritage Convention, yet this has been an important field of research for many decades.

One issue that further work on the relationship between biodiversity and cultural diversity will help address is that of the limitations associated with the 'outstanding value' approach adopted and applied in the context of the World Heritage Convention. In fact, experiences have demonstrated that even sites culturally less deserving (because less aesthetically beautiful or less 'indigenous') than others nonetheless provide ecological benefits that are crucial to the populations inhabiting them; contrariwise, sites that are ecologically simplified due to human activity may retain high cultural values according to the World Heritage criteria.[18]

What is certain is that introduction of the WHC Management Guidelines for Cultural Landscapes[19] represents a welcome evolution of the World Heritage concept/approach and one that overcomes the artificial barrier between culture and nature introduced by the conventional categories of World Heritage Sites (natural and cultural sites).

## UNESCO's Man and the Biosphere (MAB) Programme

UNESCO's Man and the Biosphere (MAB) Programme was launched in 1971—unequivocally the first international programme to espouse the concept that people and nature are inextricably linked—one year before the United Nations Stockholm Conference on the Human Environment and the creation of UNEP. MAB, essentially, operated as a research platform at the interface of people with nature. The conceptual assumption of the programme—humans (and their culture) as an integral part of the biosphere—was such that the research programme in MAB did not make a clear distinction between the natural and social sciences. Among the originally conceived MAB project areas, most anticipated today's international debate on diversity and its various forms, and the relationship between cultural and biological diversity.

Theoretical work was complemented with place-based action. The World Network of Biosphere Reserves today has close to 500 sites in 102 countries. Biosphere Reserves are landscapes or seascapes, with their set of issues, problems and opportunities both in ecological and in socio-cultural terms. These sites are learning laboratories for testing ideas and developing pragmatic and site-specific solutions to resolving problems at the human–ecosystem boundary. After 35 years of existence, MAB has developed a relatively complex bureaucratic process, typical of large institutions made up of national committees, programmes, advisory bodies, expert committees, etc. Its main message has been incorporated into the global development agenda (people as part of nature, and nature as a key element of the development agenda), but the programme's scientific capacity to further elucidate the relationship between culture and nature appears largely lost.

Hope is with recent developments within MAB on the use of biosphere reserves in promoting the conservation and sustainable use of genetic resources based on traditional knowledge and in the context of agricultural and aqua-cultural practices, together with the International Plant Genetic Resources Institute; and recently undertaken work on *terroirs*—pieces of land that do not follow administrative boundaries and which are delimited by cultural identity, namely culinary practices and language that normally reflect specific edaphic (climate and soil) properties.

## CBD: The Main Focus for Indigenous Issues and Natural Heritage

The Convention on Biological Diversity (CBD)[20] is one of the major intergovernmental processes on the environment, which recognises the dependency of indigenous and local communities on biological diversity and the unique role of indigenous and local communities in managing biodiversity—and so involved also, we believe, with intangible heritage.

This recognition is enshrined in the Convention text. Specifically, Article 8(j) calls on Parties to respect, preserve and maintain the knowledge, innovations and practices of indigenous and local communities relevant for the conservation of biological diversity, to promote their wider applications with the approval of knowledge holders and to encourage equitable sharing of benefits arising out of the use of biological diversity.

For millennia, indigenous peoples have managed their traditional lands and waters and the diversity of life contained therein and must continue to do so if indigenous peoples and Parties to the CBD are to achieve their mutual goals of conservation of biodiversity, sustainable use of its components and equitable sharing of benefits arising from its use. These three goals are also the cornerstones of indigenous societies.

The CBD, the Convention on Cultural Expressions and its preceding Declaration on Cultural Diversity share a similar value-based system and theoretical scaffolding on which a deeper understanding of cultural and biological diversity rests and should be further strengthened to foster their ultimate common principles. Important linkages also exist as to the social, economic, educational, cultural, recreational and aesthetic values of biological diversity and its components. This is being pursued with vigour through the work programme of Article 8(j) and throughout the thematic areas of the Convention.

The work of the Convention on, *inter alia*, protected areas and sacred sites, or on indicators towards achieving the 2010 Biodiversity Target offer clear indications of the co-dependence of different forms of diversity.[21] The 2010 target has an indicator of linguistic diversity and numbers of speakers of indigenous languages (which often overlaps with regions of high biodiversity). Indeed, similar to the extinction crisis of the planet's species and genetic variety, the variety and wealth of languages—and associated cultural traits—are in danger of extinction.

Beyond their intrinsic value, plants, animals and ecosystems, in their variety and distinctness, present unique emotional and physical benefits to our lives and play an integral part in culture. Their loss, which equates to the loss of diversity within and among human civilisations, impoverishes us immensely. Promotion and protection jointly of biological and of cultural diversity under the CBD is thus a unique opportunity to unite and co-operate.

Ultimately, the conservation, sustainable use, the fair and equitable access to benefits deriving from biological diversity and the protection and promotion of the diversity of cultural expressions will strengthen friendly relations among states, can assist in attaining the Millennium Development Goals, and can contribute to peace for humankind. Contrariwise, the development of positive relations between cultures can occur only in healthy environments.

At the time before time began, as indigenous peoples might say, they started a conversation with nature that continues to this very day—it is the basis of an oral contract of mutual obligation—the breaking of which has dire consequences for all of humanity.[22]

*Ramsar: The Increasing Importance of Culture in Convention Discussions—Culture and the Water Debate*

It is an imperative that a convention dealing with water and wetlands must have a strong cultural component. And yet the discussions on the role or not of culture in the deliberations of the Convention have been rather uninformed and too frequently divisive. The reasons for this are interesting to analyse. Certainly it is not due to lack of mandate, as the Convention text explicitly states that the Contracting parties *BEING CONVINCED that wetlands constitute a resource of great economic, cultural, scientific, and recreational value, the loss of which would be irreparable.*[23]

In other words, the original parties negotiating the Convention in 1971 recognised culture as one of the imperatives to be taken into account. But it received scant attention until the seventh Conference of the Parties (CoP), which had the theme 'People and Wetlands: the Vital Link'. A key discussion here involved the adoption of a resolution[24] on participatory management, focusing on the role of indigenous and local communities in wetland management.

This led to the eighth CoP, which was themed as 'Wetlands: Water, Life, and Culture'. Some very specific materials were presented for discussion and decision, including considerable preparatory material. The CoP itself was the source of a book on culture and wetlands,[25] as well as an exhibition. Resolution VIII.19[26] dealt with the issue of taking into account cultural values, but in its operative part merely 'TAKES NOTE WITH INTEREST of the list of *Guiding Principles* included in the Annex to this Resolution'.

This resolution had the following general principle: *There is a strong link between wetland conservation and benefits to people. In addition, a positive correlation between conservation and the sustainable use of wetlands has been repeatedly demonstrated. Therefore, conservation requires the involvement of indigenous peoples and local communities and cultural values offer excellent opportunities for this.*

The 23 guiding principles which followed this general principle are a skeletal version of the draft resolution, which was the subject of very vigorous debate at CoP8. What was interesting was the polarisation this debate revealed between regions of the Convention, and even within regions. Part of the reason for the polarisation was the discomfort that some parties found of dealing with culture in a convention they felt should only deal with material (*sic*) issues, and part that cultural considerations could open up the Convention more broadly than its remit allows. The result was that cultural values in wetland identification and management were sidelined.

The ninth CoP had the theme 'Wetlands and Water: Supporting Life, Sustaining Livelihoods', and was less culturally focused than the two previous CoPs but certainly continued the people–wetland interaction. Taking up the threads of the discussions at CoP8, several technical presentations attempted to devise a suitable way forward for the discussions. What emerged from the discussions was continuing discomfort with the idea of using culture as a primary reason for *identification* of wetlands of international importance. Parties, however, agreed to a new resolution, which *inter alia* established a working group on the issue, which is expected to produce a new draft decision for consideration at CoP10 in 2008.

A clear conclusion is that the resolute focus on culture as part of the listing process of the Convention, rather than the contribution of cultural and traditional knowledge to wetland management, has been detrimental to the implementation of the cultural aspects of the Convention.

*CITES: Culture, Conservation and Trade Issues*

CITES, while a Convention rooted in the conservation of biodiversity, is also one dealing extensively with trade issues, yet has not been exposed much to debates on the specific issue of culture.[27] Decision making in the Convention has usually focused on scientific and conservation concerns, based on advice from strong science-based committees. On the other hand the process of trade itself is one that has cultural origins and cultural trappings, and one might expect some resonance in discussions of the Convention.

The decision at CITES Conference of the Parties 13 (CoP13) to allow non-commercial trade in individually marked and certified *ekipas*[28] incorporated in finished jewellery for non-commercial purposes from Namibia is a recent one where cultural considerations played a role. The government of Namibia submitted a note to CoP13[29] which said, in part:

> The production of high-value modern jewellery pieces containing traditional ivory amulets, known as *ekipas* is well-established in Namibia. Such items have thus far used antique *ekipas* considered as pre-Convention ivory.
>
> *Much of Namibia's cultural heritage has been lost* through the export of such pieces, and it is evident that the supply of antique *ekipas* has become severely limited. The Ministry of Environment and Tourism (MET) has accordingly designed a system for the legal production of new *ekipas* in collaboration with the jewellery industry of Namibia. (Authors' emphasis)

*Ekipas* are unique cultural objects (round or oval carved ivory objects with a traditional geometric design, originally used for purposes of status, culture and barter) found only in northern Namibia and southern Angola, and have become much in demand because of their aesthetic quality and cultural-historic value, and as elements in modern jewellery. Many *ekipas* have been exported as pre-Convention specimens.

The CITES CoP did agree to allow this trade—essentially allowing the Convention to operate exceptionally in the pursuit of cultural objectives. But the CoP discussion was *not*, however, primarily about the cultural aspects, but rather whether the populations concerned can sustain harvesting, or other related activities. However, the fact that scientific discussions occurred was a result of cultural imperatives.

*International Whaling Commission: The Cultural Context of the Most Divisive Global Conservation Debate*

Among the many legal regimes that deal in various ways with whaling, the most significant is the 1946 International Convention for the Regulation of Whaling (ICRW),[30] of which the governing body is the International Whaling Commission (IWC). With regard to the themes explored in this paper, the IWC offers a striking example of the increasingly complex and indeterminate nature of contemporary attitudes towards cultural and biological diversity—and heritage.[31]

The ICRW is at some variance with the current forms of ecological concern and knowledge, and now stands in an uneasy relationship to the diverse cultural contexts within which whaling—as tradition to be preserved, abomination to be outlawed, or environmental challenge to be managed—is set. While legal whaling regimes are needed to ensure adequate and appropriate conservation and management, there is a matrix of cultural values which form the worldviews[32] for people and their relationship with whales.

The close links between sustainable use of wildlife, the rights of indigenous peoples and the issue of genetic resources have long been the subject of discussion in the literature.[33] Both Inuit and scientists have recognised that whale populations that are hunted on a sustainable basis have less disease, more food, and reproduce faster than whale populations not hunted sustainably. Whaling is exemplary of issues that bring into play scales from the most local to the radically global, and it is not incidental that considerations of the interests of indigenous peoples raise the question 'diversity for whom?' It is striking, however, that most debate is set in the context of Western supermarkets and coffee bars, and takes little account of the concerns of whaling communities and their cultural contexts.

Confusion between the original aims of the Convention and its modern interpretation remains a key issue. Many parties are now interpreting the Convention much more as a conservation instrument in tune with the current global environmental ethic. Restrictions imposed in implementing this approach are not, however, universally agreed or accepted by some of the communities most affected. Thus at the heart of the international regulation of whaling lies a clash of cultures with hunting communities concerning the responsible use of the resource, which currently plays out as the ascendancy of global orthodoxy over cultural imperatives.

### UNCCD: Cultural Aspects of Combating Desertification

From the outset the UNCCD has acknowledged the important role of cultural values and cultural diversity in combating desertification. It has embraced traditional knowledge, as part of the complex of cultural diversity, as a way of assisting local communities to respond to the global problem of desertification.

A tenet of the UNCCD is that 'desertification is a global problem with local solutions'.[34] From its inception, the UNCCD strategy was to build upon traditional technology, know-how and practices with the aim of increasing the ability of both government and stakeholders to control agricultural risk by improving techniques and restoring degraded lands. The leitmotif for the Convention is that 'People who bear the brunt of desertification, and who best understand the fragile environments in which they live, must be the starting point for efforts to rehabilitate desertification and combat land degradation.'[35]

The UNCCD has seven foci of action where cultural diversity and cultural values play a role:

- combating of wind and water erosion;
- hydraulic organisation for water conservation;

- improvement of soil fertility;
- vegetation protection;
- forestry;
- social organisation;
- architecture and energy.

Each cultural practice is not an expedient to solve a single problem but is an elaborated and often multipurpose system that is part of an integral approach (society, culture and economy) which is strictly linked to an idea of the world based on the careful management of local resources.

## Future Developments and Prospects: The Role of Cultural Landscapes

As we noted above, defining cultural landscapes is important for illustrating the potential of biodiversity and cultural diversity interactions for the conservation and sustainable use of biodiversity, and the resilience of cultures and societies. Frequently this means leaving intact elements of landscape connectivity created by cultural activities, or restoring such connectivity where lost. Often the research agenda has achieved more on the form, function and evolution of cultural landscapes than normative instruments.[36] And essentially all landscapes are subject to cultural influences and, as such, maintenance of ecosystem services and conservation of biological diversity are achievable only when cultural diversity is maintained. Our management of biodiversity thus becomes a cultural expression, and, in turn, biodiversity reshapes human culture.[37]

The intimate link between science, culture, socio-economic concerns and sustainable development must be strengthened.[38] Defining the research agenda on the inter-linkages between biodiversity and cultural diversity is a sine qua non for the debate in this area to continue. Without a sound research agenda, funding organisations will not be willing to incorporate the cultural element into the environmental development agenda. Moreover, institutions dealing with cultural diversity and indigenous people, protection of cultural heritage in sites where access to resources and biodiversity is required, environmental aspects of cultural tourism, culture and environment (e.g. cultural heritage, cultures and climate change), etc. need to operate on an informed basis so as to take into due account the ecological and biodiversity aspects of those issues. Pure research, however, is not the only way forward; research needs to be informed, and tuned to monitoring as well as to enlarging our understanding of the natural world and our place in it.

Prioritising specific activities of the programmes of work of international organisations and multilateral environmental agreements that relate to the dialectic between biodiversity and cultural diversity is also a key challenge. The recently established Biodiversity Liaison Group[39] is one mechanism that can help here.

Some key areas of new or reinforced research activity include:

- examining the definition of cultural landscapes, and linking these definitions back to an analysis of the multi-functional nature of landscapes;

- development of a value-based system and related theoretical basis on which a deeper understanding of interlinkages between cultural and biological diversity can be developed;
- the role of languages as forms of knowledge for understanding patterns in the maintenance of the Earth's diverse ecosystems;
- the testing of the Ramsar Guiding Principles on Wetlands and culture in wetland and other contexts;
- research on the multidimensional (multipurpose) nature of each cultural practice affecting biodiversity;
- research against the basic assumption that essentially all landscapes are subject to cultural influences, and, as such, maintenance of ecological services and conservation of biological diversity are achievable only when cultural diversity is maintained;
- verifying the hypothesis that current economic systems act as tools for social desegregation and do not adapt rapidly to new ecological and sociological knowledge, but rather mainly to new market knowledge.

These seven areas of research are needed to help define more clearly the respective roles of culture and nature, in the context of diversity, and then to set those roles in the context of developing wise and appropriate policy for sustainable living at national, regional and international levels.

## Notes

[1]  With apologies to Lewis Carroll.
[2]  'Indigenous peoples' is a term that is often misunderstood, or sometimes seen as provocative! For our purposes we use the careful definition provided in DESA, 2004. This definition is as follows:

> Indigenous communities, peoples and nations are those which, having a historical continuity with pre-invasion and pre-colonial societies that developed on their territories, consider themselves distinct from other sectors of the societies now prevailing on those territories, or parts of them. They form at present non-dominant sectors of society and are determined to preserve, develop and transmit to future generations their ancestral territories, and their ethnic identity, as the basis of their continued existence as peoples, in accordance with their own cultural patterns, social institutions and legal system.

> This historical continuity may consist of the continuation, for an extended period reaching into the present of one or more of the following factors:

> a)  Occupation of ancestral lands, or at least of part of them;
> b)  Common ancestry with the original occupants of these lands;
> c)  Culture in general, or in specific manifestations (such as religion, living under a tribal system, membership of an indigenous community, dress, means of livelihood, lifestyle, etc.);
> d)  Language (whether used as the only language, as mother-tongue, as the habitual means of communication at home or in the family, or as the main, preferred, habitual, general or normal language);
> e)  Residence on certain parts of the country, or in certain regions of the world;
> f)  Other relevant factors.

On an individual basis, an indigenous person is one who belongs to these indigenous populations through self-identification as indigenous (group consciousness) and is recognized and accepted by these populations as one of its members (acceptance by the group).

This preserves for these communities the sovereign right and power to decide who belongs to them, without external interference.

[3]    See Bennett, *Linkages in the Landscape.*

[4]    Oviedo and Maffi, *Indigenous and Traditional Peoples of the World and Ecoregion Conservation.*

[5]    See de Cuéllar, *Our Creative Diversity*; and Bridgewater and Bridgewater, 'Is there a Future for Cultural Landscapes?'

[6]    See Posey, *Cultural and Spiritual Values of Biodiversity*; and more recently Deil et al., 'Sacred Groves in Morocco'.

[7]    See Posey, *Cultural and Spiritual Values of Biodiversity.*

[8]    Governing Council resolution 22/16 [accessed 18 August 2006], available from http://www.unep.org/gc/gc22/

[9]    http://www.eblida.org/topics/wto/unesco_cultdiv.pdf [accessed 18 August 2006].

[10]   See Vitousek et al., 'Human Domination of Earth's Ecosystems'; and also Millennium Ecosystem Assessment, *Ecosystems and Human Well-being.*

[11]   See Posey, *Cultural and Spiritual Values of Biodiversity.*

[12]   Ibid.

[13]   The Human Rights Council adopted Resolution 2006/2, titled 'Working group of the Commission on Human Rights to elaborate a draft declaration in accordance with paragraph 5 of the General Assembly resolution 49/214 of 23 December 1994', on 29 June 2006 and forwarded the Declaration to the General Assembly for possible adoption before the end of 2006.

[14]   http://www.un.org/millennium/declaration/ares552e.htm [accessed 18 August 2006].

[15]   http://www.unesco.org/mab/mabProg.shtml [accessed 18 August 2006].

[16]   http://whc.unesco.org/ [accessed 18 August 2006].

[17]   http://whc.unesco.org/en/list/421 [accessed 18 August 2006].

[18]   See Hobbs et al., 'Novel Ecosystems'.

[19]   In the Operational Guidelines [accessed 18 August 2006], available from http://whc.unesco.org/archive/opguide05-en.pdf

[20]   http://www.biodiv.org/default.shtml [accessed 18 August 2006].

[21]   http://www.biodiv.org/2010-target/default.shtml [accessed 18 August 2006].

[22]   Inspired by Serres, *The Natural Contract.*

[23]   http://ramsar.org/key_conv_e.htm [accessed 18 August 2006].

[24]   http://www.ramsar.org/res/key_res_vii.08e.htm [accessed 18 August 2006].

[25]   See MIMAM.

[26]   http://www.ramsar.org/res/key_res_viii_19_e.htm [accessed 18 August 2006].

[27]   Pers. comm., Stephen Nash, CITES Secretariat.

[28]   Ivory trinkets or traditional ivory amulets.

[29]   http://www.cites.org/common/cop/13/inf/E13i-33.pdf#search=%22ekipas%22 [accessed 18 August 2006].

[30]   http://www.iwcoffice.org/commission/convention.htm [accessed 18 August 2006].

[31]   See Bridgewater, 'Whales and Wailing'.

[32]   See Bridgewater and Bridgewater, 'Is there a Future for Cultural Landscapes?'

[33]   Lynge, *The Story of the Inuit Circumpolar Conference.*

[34]   http://www.unccd.int [accessed 18 August 2006].

[35]   http://www.unccd.int//publicinfo/publications/docs/traditional_knowledge.pdf?bcsi_scan_EC783A0C3C997A81=0&bcsi_scan_filename=traditional_knowledge.pdf [accessed 18 August 2006].

[36]    See Bennett, *Linkages in the Landscape*; Bridgewater and Arico, Conserving and Managing Biodiversity Sustainably'; Jongman, *The New Dimensions of the European Landscape*.

[37]    See de Cuéllar, *Our Creative Diversity*.

[38]    See Posey, *Cultural and Spiritual Values of Biodiversity*; Bérard et al., *Biodiversity and Local Ecological Knowledge in France*.

[39]    http://www.biodiv.org/cooperation/related-conventions/blg.shtml    [accessed    18    August 2006].

# References

Bennett, A. F. *Linkages in the Landscape*. 2nd ed. Gland: IUCN, 2003.

Bérard, L., M. Cegarra, M. Djama, S. Louafi, P. Marchenay, B. Roussel and F. Verdeaux. *Biodiversity and Local Ecological Knowledge in France*. Paris: CIRAD, 2005.

Bridgewater, P. B. 'Whales and Wailing'. *International Social Science Journal* 55 (2003): 555–59.

Bridgewater, P. B. and S. Arico. 'Conserving and Managing Biodiversity Sustainably: The Roles of Science and Society'. *Natural Resources Forum* 26 (2002): 245–48.

Bridgewater, P. B. and C. Bridgewater. 'Is there a Future for Cultural Landscapes?' In *The New Dimensions of the European Landscape,* edited by R. H. G. Jongman. Berlin: Springer, 2004.

Cuéllar, J. P. de. *Our Creative Diversity—Report of the World Commission on Culture and Development*. Paris: UNESCO, 1995.

Deil, U., H. Culmsee and M. Berriane. 'Sacred Groves in Morocco: A Society's Conservation of Nature for Spiritual Reasons'. *Silva Carelica* 49 (2005): 185–201.

DESA Workshop on Data Collection and Disaggregation for Indigenous Peoples, New York, 19–21 January 2004, PFII/2004/WS.1/3, United Nations, New York, 2004.

Hobbs, R. J., A. Salvatore, J. Aronson, J. S. Baron, P. Bridgewater, V. A. Cramer, P. R. Epstein, J. J. Ewel, C. A. Klink, A. E. Lugo, D. Norton, D. Ojima, D. M. Richardson, E. W. Sanderson, F. Valladares, M. Vilà, R. Zamora and M. Zobel. 'Novel Ecosystems: Theoretical and Management Aspects of the New Ecological World Order'. *Global Ecology and Biogeography* 15 (2006): 1–7.

Honari, M. and T. Boleyn. *Health Ecology: Health, Culture and Human Environment Interaction*. London: Routledge, 1999.

Jongman, R. H. G., ed. *The New Dimensions of the European Landscape*. Berlin: Springer, 2004.

Lynge, A. *The Story of the Inuit Circumpolar Conference*. Greenland: ICC, Nuuk, 1993.

Millennium Ecosystem Assessment. *Ecosystems and Human Well-being*. Washington, DC: Island Press, 2005.

MIMAM. *El patrimonio cultural de los humedales* (Wetlands cultural heritage). Madrid: Dirección General de Conservación de la Naturaleza, 2002.

Oviedo, G. and L. Maffi. *Indigenous and Traditional Peoples of the World and Ecoregion Conservation: An Integrated Approach to Conserving the World's Biological and Cultural Diversity*. Gland: WWF International-Terralingua, 2000.

Posey, D. A. *Cultural and Spiritual Values of Biodiversity*. Nairobi: ITP-UNEP, 1999.

Serres, M. *The Natural Contract*. Translated by E. MacArthur and W. Paulson. Ann Arbor: University of Michigan Press, 2003.

Vitousek, P. M., H. A. Mooney, J. Lubchenco and J. Melillo. 'Human Domination of Earth's Ecosystems'. *Science* 277 (1997): 494–99.

# The Gift of Environment: Divine Response and Human Responsibility

## Bartholomew
*Archbishop of Constantinople and Ecumenical Patriarch*

Thine own from Thine own we offer to you, in all and through all.

St John Chrysostom (347–405)

### Heritage, Gift, and Liturgy

As we contemplate the title of your journal and respond to the gracious invitation to contribute certain remarks on our concern for the protection of the natural environment, it becomes immediately apparent that the word 'heritage' in many ways crystallises the very essence of the Orthodox theological and spiritual approach towards creation. For the term 'heritage' indicates the fact that the world we enjoy comprises a gift we have received and not some property we own. Moreover, something inherited must be preserved and conveyed to generations that succeed us, and not wasted selfishly without concern for those who come after us. Therefore, heritage' is accompanied by a sense of respect and responsibility, which resembles the notion of tradition in religious circles.

Just as the word 'heritage' bears a significant and symbolic dimension, so too the word 'ecology' contains the prefix 'eco', which derives from the Greek word *oikos*, signifying 'home' or 'dwelling'. How unfortunate, then, and indeed how selfish it is that we have reduced the meaning and restricted the application of this crucial word. For this world is indeed our home. Yet it is also the home of everyone, as it is the home of every creature, as well as of every form of life created by God. It is a sign of arrogance to presume that we human beings alone inhabit this world. Moreover, it is a sign of arrogance to imagine that only the present generation inhabits this Earth.

The above words from the Divine Liturgy attributed to our 4th-century predecessor and Archbishop of Constantinople, St John Chrysostom, symbolise our conviction that this world is the fruit of divine generosity and boundless grace as well as of our commitment to respond with gratitude by respecting and protecting the natural environment—or, as we are commanded in Scripture, by 'serving and preserving the earth' (Gen. 2.15) for the sake of future generations.

## Creation as Gift from God

'Gift' (*doron*, in Greek) and 'gift-in-return' (*antidoron*, in Greek) are liturgical terms that define the Orthodox theological understanding of the environmental question in a concise and clear manner. On the one hand, the natural environment comprises the unique *doron* of the Triune God to humankind. On the other, the appropriate *antidoron* of humankind towards its divine Maker is precisely the respect for and preservation of this gift, as well as its responsible and proper use. Each believer is called to celebrate life in a way that reflects the words of the Divine Liturgy: 'Thine own from Thine own we offer to you, in all and through all.'

Thus the Eastern Orthodox Church proposes a liturgical worldview. It proclaims a world richly imbued by God, and a God profoundly involved in this world. Our original sin, so it might be said, does not lie in any legalistic transgression that might incur divine wrath or human guilt. Instead, it lies in our stubborn refusal as human beings to receive the world as a gift of reconciliation with our planet and to regard the world as a sacrament of communion with the rest of humanity.

This is the reason why the Ecumenical Patriarchate has initiated and organised a number of international and interdisciplinary symposia over the last decade: on the Aegean Sea (1995), the Black Sea (1997), the Danube River (1999), the Adriatic Sea (2002), the Baltic Sea (2003) and, most recently, the Amazon River (2006).[1] For, like the air that we breathe, water is the very source of life; if defiled or despoiled, the element and essence of our existence is threatened. Put simply, environmental degradation and destruction is tantamount to suicide. One of the hymns of the Orthodox Church, chanted on the day of Christ's Baptism in the River Jordan, a feast of renewal and regeneration for the entire world, states:

> I have become … the defilement of the air and the land and the water.

At a time when we have polluted the air that we breathe and the water that we drink, we are called to restore within ourselves a sense of awe and delight, to respond to matter as to a mystery of ever-increasing connections.

As a gift from God to humanity, creation becomes our companion, given to us for the sake of living in harmony with it and with others. We are to use its resources in measure, to cultivate it in love, and to preserve it in accordance with Scriptural command (cf. Gen. 2.15). Within the unimpaired natural environment, humanity discovers deep spiritual peace and rest; and in humanity that is spiritually cultivated by the peaceful grace of God, nature recognises its harmonious and rightful place.

Nevertheless, the first-created human being misused freedom, preferring alienation from God-the-Giver and attachment to God's gift. Consequently, the double relationship of humanity to God and creation was distorted and humanity became preoccupied with using and consuming the Earth's resources. In this way, the human blessedness derived from the love between God and humanity ceased, and humanity sought to fill this void by drawing from creation the blessedness that was lacking. From grateful user, the human person became greedy abuser. In order to remedy this situation, human beings are called to be 'eucharistic' and 'ascetic', namely to be thankful by offering glory to God for the gift of creation, while at the same time being respectful by practising responsibility in the web of creation.

**Eucharistic and Ascetic Beings**

Let us reflect further on these two words 'eucharistic' and 'ascetic'. The implications of the first word are quite easily appreciated. In calling for a 'eucharistic spirit', the Orthodox Church is reminding us that the created world is not simply our possession but rather is a gift—a gift from God the Creator, a healing gift, a gift of wonder and beauty. Therefore, the proper response, upon receiving such a gift, is to accept and embrace it with gratitude and thanksgiving. This is surely a distinctive characteristic of human beings. Humankind is not merely a logical or a political animal. Above all, human beings are eucharistic animals, capable of gratitude and endowed with the power to bless God for the gift of creation. Other animals express their gratefulness simply by being themselves, by living in the world through their own instinctive manner; yet we human beings possess self-awareness in an intuitive manner, and so consciously and by deliberate choice we can thank God with eucharistic joy. Without such thanksgiving, we are not truly human.

A eucharistic spirit implies using the Earth's natural resources with thankfulness, offering them back to God; indeed, not only them but also ourselves. In the Sacrament of the Eucharist, we return to God what is His own, namely the bread and the wine, together with the entire community. All of us and all things represent the fruits of creation, which are no longer imprisoned by a fallen world, but returned as liberated, purified from their fallen state, and capable of receiving the divine presence within themselves. Whoever gives thanks also experiences the joy that comes from the appreciation of that for which he or she is thankful. Conversely, whoever does not feel the need to be thankful for the wonder and beauty of the world, but instead demonstrates only selfishness or indifference, can never experience a deeper, divine joy, but only sullen and inhumane satisfaction.

Second, we have the 'ascetic ethos' of Orthodoxy that involves fasting and other similar spiritual disciplines. These make us recognise that everything we take for granted in fact comprises God's gifts provided in order to satisfy our needs as they are shared fairly among all people. However, they are not ours to abuse and waste simply because we have the desire to consume them or the ability to pay for them. The 'ascetic ethos' is the intention and discipline to protect the gift and to preserve nature from harm. It is the struggle for self-control, whereby we no longer wilfully consume every fruit, but instead manifest a sense of self-restraint and abstinence from certain fruits. Both the protection and the self-restraint are expressions of love for all of humanity and for the entire natural creation. Such love alone can protect the world from unnecessary waste and inevitable destruction. After all, just as the true nature of God is love (1 John 4.8), so too humanity is originally and innately endowed with love.

Our purpose is thus enjoined to the priest's prayer in the Divine Liturgy: 'In offering to Thee, Thine own from Thine own, in all and through all—we praise Thee, we bless Thee, and we give thanks to Thee, O Lord.' Then, we are able to embrace all people and all things—not with fear or necessity, but with love and joy. Then, we learn to care for the plants and for the animals, for the trees and for the rivers, for the mountains and for the seas, for all human beings and for the whole natural environment. Then, we discover joy—rather than inflicting sorrow—in our life and in our world. Then, we create and promote instruments of peace and life, not tools of violence and death. Then, creation on the one hand and humanity on the other hand—the one that encompasses and the one that is encompassed—correspond fully and co-operate with one another. For they are no longer in contradiction or in conflict or in competition. Then, just as humanity offers creation in an act of priestly service and sacrifice to God, so also does creation offer itself in return as a gift to humanity and to the generations that are to follow. Then, everything becomes a form of exchange, the fruit of abundance, and a fulfilment of love. Then, everything celebrates what St Maximus the Confessor in the 7th century called a 'cosmic liturgy'.

## A New Worldview

The crisis that we are facing in our world is not primarily ecological. It is a crisis concerning the way we envisage or imagine the world. We are treating our planet in an inhuman, godless manner precisely because we fail to see it as a gift inherited from above; it is our obligation to receive, respect and render this gift to future generations. Therefore, before we can effectively deal with problems of our environment, we must change the way we envisage the world. Otherwise, we are simply dealing with symptoms, not with their causes. We require a new worldview if we are to desire 'a new earth' (Rev. 21.1).

Therefore, let us acquire a 'eucharistic spirit' and an 'ascetic ethos', bearing in mind that everything in the natural world, whether great or small, has its importance within the universe and for the life of the world; nothing whatsoever is useless or contemptible. Let us regard ourselves as responsible before God for every living creature and for all the natural creation; let us treat everything with proper love and utmost care. Only in this way shall we secure a physical environment where life for the coming generations of humankind will be healthy and happy. The unquenchable greed of our generation constitutes a mortal sin inasmuch as it results in destruction and death. This greed in turn leads to the deprivation of our children's generation, in spite of our desire to bequeath to them a better future. Ultimately, it is for our children that we must perceive our every action in the world as having a direct effect upon the future of the environment.

As we declared some years ago in Venice with the late Pontiff of the Roman Catholic Church, Pope John Paul II:

> It is not too late. God's world has incredible healing powers. Within a single generation, we could steer the earth toward our children's future. Let that generation start now, with God's help and blessing.

We must frankly admit that humankind is entitled to something better than what we see around us. We and, much more, our children and future generations are entitled to a better world, a world free from degradation, violence and bloodshed, a world of generosity and love. It is selfless and sacrificial love for our children that will show us the path that we must follow into the future.

## Note

[1]   Concerning the ecological work of the Ecumenical Patriarchate, please see the following website: http://www.rsesymposia.org/

# Index